THE LOGGERS

THE LOGGERS

By the Editors of

TIME-LIFE BOOKS

with text by

Richard L. Williams

TIME-LIFE BOOKS, NEW YORK

TIME-LIFE BOOKS

Founder: Henry R. Luce 1898-1967

Editor-in-Chief: Hedley Donovan
Chairman of the Board: Andrew Heiskell
President: James R. Shepley

Vice Chairman: Roy E. Larsen

Managing Editor: Jerry Korn
Assistant Managing Editors: Ezra Bowen,
David Maness, Martin Mann
Planning Director: Oliver E. Allen
Art Director: Sheldon Cotler
Chief of Research: Beatrice T. Dobie
Director of Photography: Melvin L. Scott
Senior Text Editors: Diana Hirsh, William Frankel
Assistant Planning Director: Carlotta Kerwin
Assistant Art Director: Arnold C. Holeywell
Assistant Chief of Research: Myra Mangan

Publisher: Joan D. Manley
General Manager: John D. McSweeney
Business Manager: John Steven Maxwell
Sales Director: Carl G. Jaeger
Promotion Director: Paul R. Stewart
Public Relations Director: Nicholas Benton

THE OLD WEST

EDITORIAL STAFF FOR "THE LOGGERS"
Editor: George G. Daniels
Picture Editor: Jean Tennant
Text Editors: Valerie Moolman, Joan S. Reiter,
Gerald Simons
Designer: Bruce Blair
Staff Writers: Carol Clingan, Lee Greene,
Frank Kappler, Rosalind Stubenberg
Chief Researcher: June O. Goldberg
Researchers: Jane Coughran, Jane Jordan,
Thomas Lashnits, Donna Lucey, Nancy Miller,
Mary Kay Moran, Fred Ritchin, Vivian Stephens
Design Assistant: Joan Hoffman

EDITORIAL PRODUCTION
Production Editor: Douglas B. Graham
Assistant Production Editors:
Gennaro C. Esposito, Feliciano Madrid
Quality Director: Robert L. Young
Assistant Quality Director: James J. Cox
Associate: Serafino J. Cambareri
Copy Staff: Eleanore W. Karsten (chief),
Barbara H. Fuller, Gregory Weed,
Florence Keith, Pearl Sverdlin
Picture Department: Dolores A. Littles,
Marianne Dowell
Traffic: Carmen McLellan

THE AUTHOR: Seattle-born and raised, Richard L. Williams was on home ground in writing this volume, his third about the Northwest for TIME-LIFE BOOKS; he also authored *The Northwest Coast* and *The Cascades* volumes in The American Wilderness series. He has been a writer, editor and correspondent for the *Seattle Times*, Dell Publishing Company, and TIME and LIFE magazines. As a member of the editorial staff of TIME-LIFE BOOKS, he was editor of the Foods of the World series and the LIFE Library of Photography. In 1975 he joined the staff of *Smithsonian* magazine.

THE COVER: Perched on the felled trunk of a great redwood, two loggers begin the long, laborious — and often wickedly hazardous — task of making it ready for market. In this 1876 engraving (to which color has been added) the sawyer, or bucker, prepares to cut the log into sections, while the peeler stands by with the bar he uses to pry off the bark for easier sawing; the team of oxen, foreground, waits to haul away the bucker's cuts. In the frontispiece photograph, two fallers balance nonchalantly on their springboards after completing the undercut in a Douglas fir. The logger on the left holds a bottle of oil used to lubricate the saw.

Valuable assistance was provided by the following departments and individuals of Time Inc.: Editorial Production, Norman Airey; Library, Benjamin Lightman; Picture Collection, Doris O'Neil; Photographic Laboratory, George Karas; TIME-LIFE News Service, Murray J. Gart.

CONTENTS

Clouds of steam from power saws rise in the late 1860s over the Albion lumber mill, located 115 miles north of the San Francisco area.

1 | A treasure richer than gold

Awhirl with the racketing activities of its booming new industry, the Albion lumber mill, on California's Mendocino County coast, typified the logging outposts that were bringing to a still-incredulous world the prodigious treasure of the Far Western forests. It had all the requisites: a dry, flat point of land near a river and close enough to deep water to load schooners; hillsides sloping steeply to the river, the better to slide logs to the water — and, of course, unparalleled stands of conifers.

Here the great evergreens were principally coast redwoods. Inland and to the north, the rugged land was covered with Douglas fir, Sitka spruce, sugar pine and western redcedar. Against their colossal boles, loggers from the depleted forests back east leveled axes and saws that seemed ludicrously inadequate to the task. But with courage, tireless strength and a wide range of inventive techniques, these men, antlike in their Bunyanesque surroundings, prevailed to harvest a resource far greater than the West's fabled gold and silver.

Undaunted by the immense girth of a Sitka spruce, two turn-of-the-century "fallers" prepare to have at the giant. Their double-bitted axes were a Western-forest innovation, as was the elevated perch of planks driven into slits cut in the trunk, made necessary by the huge swelling at the base.

A team of horses begins hauling a Douglas fir out of the woods to a sawmill near Tacoma. The felled tree was pulled to this central collecting point by wire cables fastened to the stationary, steam-powered donkey engine at left.

Redwood loggers in California's Humboldt County show off a good day's work in the 1890s. This whopper measured more than 12 feet across at its base. Crews found specimens up to 18 feet thick, which could take two men a week to fell, trim and cut into manageable lengths.

A team of eight oxen drags a log to Oregon's Big Sandy Creek from the foot of a hillside in 1889. The trees have been stripped of bark, enabling them to slide on the ground more easily, but making the footing precarious for the "river pigs" at right, who kept them from jamming in the creek.

15

Loggers wade up to their waists in northern California's Klamath River to free a pine log firmly grounded on a shallow bar in the 1890s. A lesser specimen floats by serenely. The water wheel in the background, a source of inexpensive power, did not play a part in the logging operation.

"Monarchs to whom all men lift their hats"

In the summer of 1869, Samuel Wilkeson, a business agent for the Northern Pacific Railroad, was sent west to inspect the company's proposed rail route from Minnesota into the sparsely settled wilderness of Washington Territory. Wilkeson had a good idea of what he would find out there; the Northwest country was already world-famous for its prodigious forests, giant trees and superb lumber. Nevertheless, when Wilkeson saw those woodlands for himself, he was thunderstruck.

"Oh! What timber," he wrote in his report. "These trees—these forests of trees—so enchain the sense of the grand and so enchant the sense of the beautiful that I linger on the theme and am loth to depart. Forests in which you cannot ride a horse—in which you cannot possibly recover game you have shot without the help of a good retriever—forests into which you cannot see, and which are almost dark under a bright midday sun —such forests, containing firs, cedars, pine, spruce and hemlock, envelop Puget Sound and cover a large part of Washington Territory, surpassing the woods of all the rest of the globe in the size, quantity and quality of the timber."

Wilkeson was able to support his grandiose assessment with cold facts. The great trees—he called them "monarchs to whom all worshipful men inevitably lift their hats"—yielded an incredible five times more timber per acre than the trees of the Eastern forests. He reported that in the year 1869 alone the 14 sawmills in the Puget Sound area had disgorged "over 170 million board feet" (a board foot measured one foot by one foot by one inch). Furthermore, after decades of commercial logging, the Northwestern forests, "covering hundreds and hundreds of square miles," were "as yet scarcely scarred."

At that, Wilkeson underestimated the woodlands. They covered not hundreds and hundreds of square miles but thousands and thousands of square miles—the greatest concentration of the greatest trees known to man *(page 20)*. The Pacific Slope forests alone, stretching from Canada far down into California, formed a dense green belt which was 1,200 miles long, 30 to 150 miles wide and 140,000 square miles in area; they blanketed both sides of the rugged Coast Ranges and, farther inland, climbed the flanks of the Cascade mountains and the Sierra Nevada. Still farther inland, another broad band of forests—which covered 170,000 square miles—ran along the slopes of the Rocky Mountains. And between the two immense forest chains, lesser woodlands enveloped sections of valleys and plains.

Just as these Western forests were vast beyond comprehension, so were they rich beyond calculation—a treasure surpassing all the West's great gold and silver strikes that inflamed the American imagination. The timber was more prosaic, perhaps, than the glittering minerals, but it brought enormous wealth and power to the men who harvested the trees, who converted the logs to lumber and who did the work of moving the lumber to the marketplace.

The exploitations of those colossal forests began in the late 1820s, and the pioneer period of Western logging lasted for about nine decades, gaining strength and momentum year after year. Tens of thousands of men went west to profit from the timber harvest, whether by felling trees or building a corporative empire. But many of them came for something more. The timberlands of the West offered a challenge, an adventure, an unencumbered way of life, a chance to be part of a

---

A stand of California redwoods looms over a clearing in this 1890s photograph by A. W. Ericson, who stood his child beside one of the behemoths to dramatize its size.

new and important endeavor with an exciting future.

The early enterprisers were hard-headed realists, but visionaries nonetheless. In their ranks were small sawmill owners from the Middle West, shipowners from New England, lumbermen who had grown up with logging in Maine and then moved on to the forests around the Great Lakes. In time, much of the woodlands fell into the hands of a few huge companies. Among them was Sam Wilkeson's Northern Pacific Railroad, which at the turn of the century owned timberlands translatable into 36 billion board feet of lumber.

The men who filled the logging gangs and ran the sawmills of the far West originally came from all over the globe. In spite of their diverse backgrounds, however, as a type they had much in common: great muscles, insatiable appetites, the daring and drive to do dangerous labor for as little as $1 a day plus board and keep, and a penchant for booze, brawls and bawds. And all were dumfounded by their first sight of the Northwest's tremendous trees. Their main stock-in-trade had been the eastern white pine, which grew to an average of 100 feet in height and almost three feet in diameter; and they had only the puniest of implements with which to tackle Western forest giants twice as tall and four or five times thicker.

But soon enough, the loggers improvised astonishing new tools and techniques for felling the trees and getting the timber out of the woods. Steam-driven "donkey" engines with steel cables a half mile long yanked immense logs to them like matchsticks. Ingeniously designed steam locomotives puffed up and down steep forested slopes, first on wooden rails, then on steel ones, carrying great cargoes of logs. Spidery, water-

A GUIDE TO THE WESTERN WOODLANDS

Of all plant life, trees need the most water; and since much of the American West was arid or semiarid, the great Western forests were found either hard by the Pacific or in the mountains, where air and soil were moist much of the time. Within those areas, distribution of species varied according to such traits as tolerance of shade and resistance to wind and cold. Some, such as the ponderosa pine and Douglas fir, could withstand a wide range of conditions and so spread over great areas. Among the most restricted were the sequoias, a family that before the great weather and soil changes of the glacial period had covered most of the hemisphere.

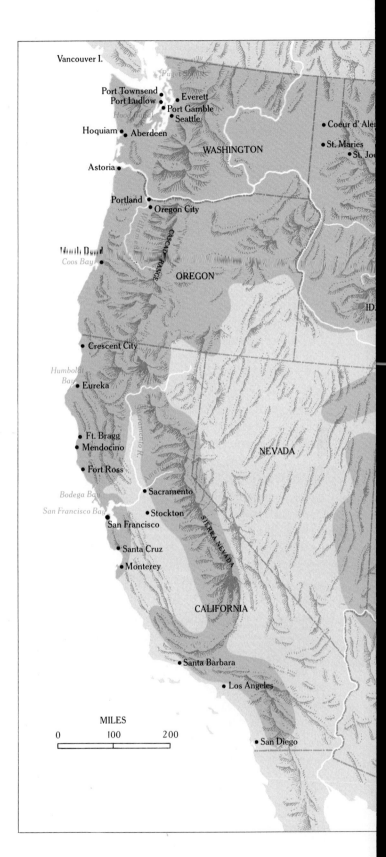

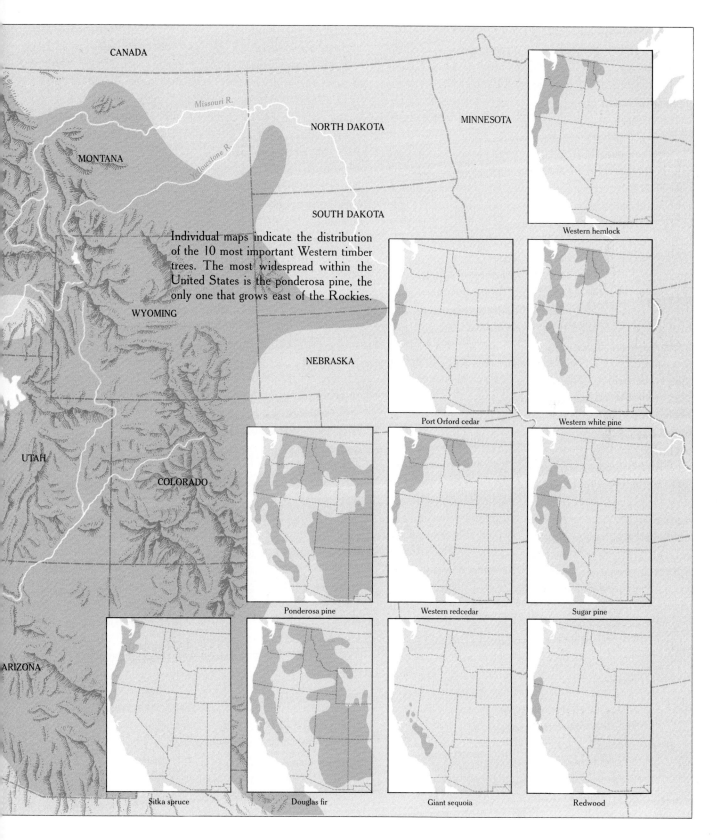

CANADA

MONTANA

Missouri R.

Yellowstone R.

NORTH DAKOTA

MINNESOTA

SOUTH DAKOTA

Individual maps indicate the distribution of the 10 most important Western timber trees. The most widespread within the United States is the ponderosa pine, the only one that grows east of the Rockies.

WYOMING

NEBRASKA

UTAH

COLORADO

ARIZONA

Western hemlock

Port Orford cedar

Western white pine

Ponderosa pine

Western redcedar

Sugar pine

Sitka spruce

Douglas fir

Giant sequoia

Redwood

filled wooden flumes snaked through the hills and ravines, hurtling logs and lumber downslope at speeds up to 50 miles an hour. The sawmill became a mass of sophisticated machinery; screaming gang saws, whining band saws and whirring circular saws jumped the daily lumber output from a few thousand board feet to hundreds of thousands.

The loggers, like the cowboys, were wilderness seminomads with no ties to home or family. Their job was to keep cutting and moving on, and they did it with remorseless efficiency. They seldom spared a tree that was worth chopping down, and they never stopped to plant one amid the graveyards of stumps they left behind. Year after year, they pushed back the forests' perimeters. They cut billions of board feet of timber for the houses and towns of the more than two million people who settled in the Far West during the second half of the 19th Century. They cut billions of board feet more for the roadbeds of the five great railroads that spanned the continent by 1893. They cut billions of board feet more for export: lumber from the Northwest built plantations in Hawaii and railroads in South America and China, shored up gold mines in Australia and diamond mines in South Africa, went into the navies and merchant ships of European nations.

All told, some 300 billion board feet of lumber were moved to market from the Western forests during the 90-year pioneer period. The terrific toll alarmed naturalists, nature lovers and government officials, who inveighed against the timber barons and demanded forest conservation. Their arguments seemed to be rebutted by a United States government report published in 1909; it showed that in spite of those nine decades of ruthless exploitation, fully 89 per cent of the Northwestern forests remained intact, still overspread with magnificent trees. The study suggested that the forests might last for many decades even if the lumbermen continued their uninhibited harvests. But it also suggested that if simple measures of conservation—selective cutting and routine replanting—were put into practice in time, the forests might last forever.

In any event, by the end of the hectic, gaudy period, the loggers and lumbermen had established Washington as the leading lumber-producing state in the nation, and the once-remote forests of the Pacific Slope and the Rockies would soon become the principal source of lumber for builders the world over. By then, the timber-based economy of the Northwest had transformed the region from a thinly settled wilderness into the fastest-developed section of the United States. And by then the Western lumber trade was fully prepared for the much greater demands that would be placed upon it in the years to come.

Thanks to nature's prodigality, it was foreordained that Americans would become great loggers—and conspicuous consumers of timber. When the first colonists landed on the Atlantic Coast in the early 17th Century, they found themselves engulfed in trees. Approximately one third of the virgin continent was heavily wooded, and the whole of the Northeast was practically a single forest, stretching 1,000 miles north and south, and extending more than 1,000 miles inland, with only a few sizable open tracts of prairie land. So dense were the coastal woodlands that for a century overland travel was an ordeal that was undertaken with the gravest of misgivings.

From the first, the colonists made lavish use of their timber riches. They cut down trees to build houses, to clear fields for crops, to burn as firewood for cooking and for winter warmth; and within a few years, they erected America's first sawmills at Jamestown in Virginia and on the Saco River in Maine. The settlers were quick to realize that timber was a valuable export, and by 1700 ships built of Maine lumber were carrying more of the same to market in the West Indies, Europe and elsewhere.

Maine was particularly well suited to large-scale commercial logging; its woodlands were enormous —they still seemed to be inexhaustible through the 18th Century—and the region's many rivers and streams were natural conveyors for bringing timber to the sawmills. Men of means and foresight in the cities of Boston, New York and Philadelphia began buying up Maine timberland at prices as low as 12½ cents an acre. One of these shrewd speculators was Philadelphian William Bingham, an influential banker and politician who in the 1790s acquired 2.1 million acres of virgin pine and spruce in the Penobscot country in back of the town of Bangor.

To cut the timber down and move it out of the forests, the landowners sent to Scotland and Ireland for

Archibald Menzies *(right),* a naturalist on
George Vancouver's 1791 expedition, is
credited with discovering the Douglas fir,
but the tree was named for a fellow Scot,
David Douglas *(left),* who conducted the
definitive study of the species in 1825.

manpower, and employed small armies of French-
Canadians, who had already learned logging in their
own dense Eastern forests. "White water logging" became the standard Maine technique. All through the
bitter cold winters, sled-loads of logs were dragged by
horses to the banks of frozen streams and stashed there
in great piles. To ease the sleds' way over forest trails,
"sprinkler men" went out by night, pouring water that
froze into solid ice by morning.

When the spring thaws came, loggers armed with
long, steel-tipped pike poles pried and pushed the piles
of logs into the streams for spectacular drives downstream, along the Penobscot River and the Kennebec
River and other waterways. The loggers took to the
rivers themselves; their logs, loose or bound into rafts,
were driven along by men in boats, by men who rode
the rafts, by men on single logs, balancing themselves
on the slippery "sticks" with their pike poles. Along
the way to the mills, the men freed "key" logs to break
up the big jams that often dammed the river, and they

were lucky if they got only dunked, not drowned.

Most of the loggers survived the drives to "blow in"
their earnings (about $20 a month) in Bangor, Skow-
hegan, East Machias and other mill towns where whiskey and women were readily available. It was there
that their hard carousing and their ready fists earned
them a fearsome reputation, which loggers never lived
down nor wanted to.

It took about a century of intensive logging to con-
firm that the forests of Maine were really not endless.
As early as the 1830s, lumbermen foresaw the ex-
haustion of the best tracts of timberland and started
looking around for new timberlands to conquer. Some
lumbermen moved their operations southward into up-
state New York and western Pennsylvania. The Erie
Canal, completed in 1825, was soon carrying loggers
westward by the thousands, and floating lumber eastward by the millions of board feet. Albany, the canal's
eastern terminus, became the busiest lumber market in
the world, and so it remained for four decades. ◉

The vital statistics of a family of giants

The monarchs of Western forests were the two varieties of redwood and the Douglas fir, but many other species dwarfed anything that grew in the East or in the Great Lakes area. The eastern white pine, tallest tree east of the Rockies, rarely exceeded 200 feet and averaged only 100; what was more important to lumbermen, the portion of the trunk clear of limbs — and therefore knots — averaged under 60 feet.

By comparison the western hemlock — smallest of the important lumber trees of the West — averaged 150 feet, with specimens sometimes soaring over 250, and its normal limb-free shaft ran to more than 70 feet. The hemlock in turn was overshadowed by both of the redwood species and the Douglas fir: all three averaged more than 200 feet in height, with specimens sometimes exceeding 300 feet. The lowest limbs of many California coast redwoods were 150 feet above the forest floor.

But it was the tremendous girths of the redwoods that made loggers' mouths water. Woodsmen measure a trunk diameter at chest height — or 4 1/2 feet above the ground. The Sierra redwood, or giant sequoia — shorter, broader cousin of the coast redwood — ran to diameters of 15 to 20 feet, and some patriarchs to nearly 40.

For all of that, the Douglas fir — though shorter than the redwoods and not so stout, with an average diameter of around 8 feet — became the greatest lumber source the world had ever known. It greatly outnumbered the redwoods and had a vast range, from British Columbia to New Mexico. Tree for tree the Douglas fir stood up very well indeed against the redwoods: the tallest conifer known to man was a Douglas fir that measured 385 feet.

Between the western hemlock and the Big Three ranged an impressive array of other huge conifers. The drawings here and on the following pages constitute a manual of the top 10

Western timber trees, with the eastern white pine included for comparison. The silhouettes have been drawn to scale on the basis of average heights, diameters and shaft lengths. The detail drawings of the cones and needles beneath each profile have not been reproduced to scale. The illustrations are the work of Charles Sprague Sargent, a Harvard botanist who was a member of the Northern Pacific Transcontinental Survey in 1882-1883 and who produced *The Silva of North America:* 14 folio volumes with definitive illustrations of every species of tree north of Mexico, between 1891 and 1902.

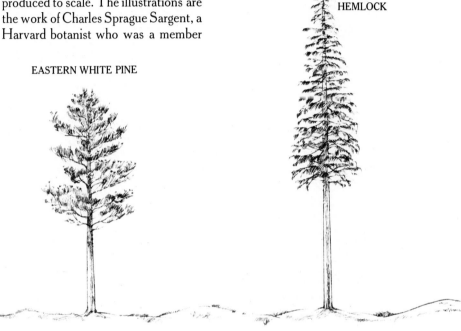

EASTERN WHITE PINE

WESTERN HEMLOCK

The most important lumber tree east of the Rocky Mountains, the eastern white pine (*Pinus strobus*) is 2 to 3 1/2 feet in diameter. Its blue-green needles are 3 to 5 inches long and mature cones are as long as 8 inches. The wood is soft and straight-grained, and weathers with little warping.

Bigger than its Eastern relative, the western hemlock (*Tsuga heterophylla*) is a moisture-loving species 2 to 4 feet in diameter. Its flat, glossy green needles are round-tipped; the cones are small, 3/4 to 1 1/2 inches long. The wood is straight-grained and relatively resistant to termites.

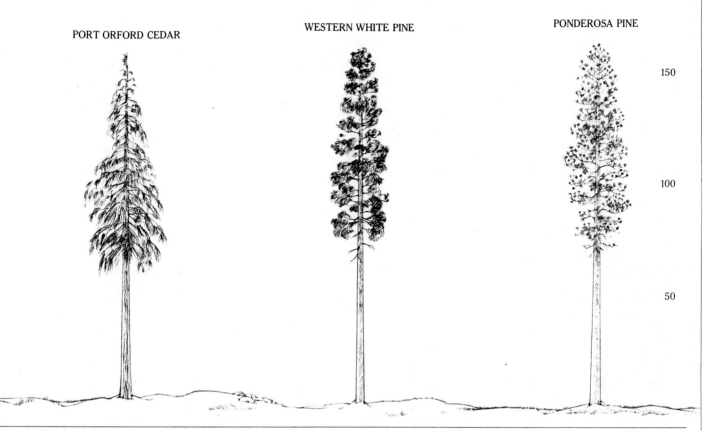

PORT ORFORD CEDAR

WESTERN WHITE PINE

PONDEROSA PINE

200 ft.

150

100

50

Overlapping bright green feathery needles about 1/16 of an inch in length clothe the branchlets of the Port Orford cedar *(Chamaecyparis lawsoniana),* found mostly in stands in southern Oregon. The wood gives off a spicy aroma that repels moths, making it ideal as a chest or closet lining.

The western or Idaho white pine *(Pinus monticola)* is a mountain dweller with a diameter of 2 to 3 1/2 feet. Its 2- to 4-inch needles are grouped in fives, and its curved 5- to 11-inch cones give it the nickname "finger-cone pine." The straight-grained wood is easily worked and weathers well.

Named for the great bulk of its 4-foot-diameter trunk, the ponderosa pine *(Pinus ponderosa)* is also called western yellow pine. Its needles, in twos and threes, are 5 to 11 inches long. The wood is fine-grained and makes excellent paneling, but it is susceptible to fungus when used outdoors.

(Continued)   25

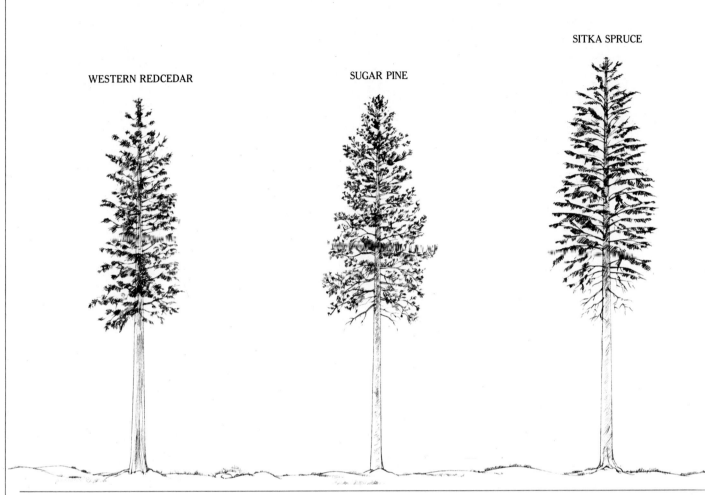

WESTERN REDCEDAR

SUGAR PINE

SITKA SPRUCE

A giant cousin of the arborvitae trees that were used in landscaping, the western red-cedar (*Thuja plicata*) is normally 6 feet in diameter. The needles are arranged in lacy sprays and the 1/2-inch cones are nearly hidden in the foliage. The wood's warp-resistance makes it classic shingle material.

Greatest of all pines, the sugar pine (*Pinus lambertiana*) is named for its sweet resin. The trunk can reach 7 feet in diameter. Its blue-green needles, grouped in fives, are 2 to 4 inches long and its cones, the largest known, run up to 2 feet in length. The wood is easily worked and resists warping.

Largest of the Northern Hemisphere's 18 spruces, the Sitka spruce (*Picea sitchensis*) rivals the Douglas fir in height but not in girth, commonly measuring 4 feet in diameter. Its prickly needles are 1/2 to 1 inch long; the cones are 2 to 4 inches in length. Thin-barked, it is vulnerable to fire.

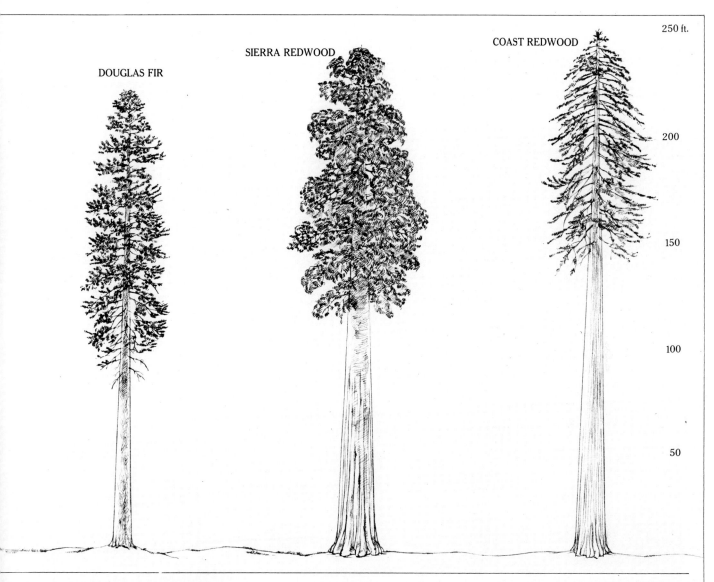

DOUGLAS FIR

SIERRA REDWOOD

COAST REDWOOD

250 ft.

200

150

100

50

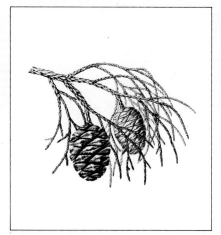

The Douglas fir *(Pseudotsuga taxifolia)* is not a true fir but a unique species of conifer. Botanists' early confusion in classifying it was a result of the tree's 1- to 1 1/2-inch needles, which grow all around the stem like spruce but are as supple as those of fir. The wood is light but strong.

The Sierra redwood or giant sequoia *(Sequoia gigantea)* is thought to have a life span of more than 5,000 years, longest of anything on earth. The sequoia's scalelike needles are similar to a cedar's and its cones are 2 to 3 inches long. The dark red wood is brittle but almost impervious to rot.

The world's tallest conifer species, the coast redwood *(Sequoia sempervirens)* has an average diameter of 13 feet. Its yellow-green needles stand out stiffly on opposite sides of the branchlet and its cones are only 1 inch long. The rot-resistance of both redwoods is attributed to high tannin content.

On his third and final voyage to the Pacific, Captain James Cook explored the Northwest coast and put into Nootka Sound on Vancouver Island in March 1778. A member of his expedition recorded this view of densely forested hillsides and Indian canoes crowding around the vessels.

Meanwhile, the logging frontier kept moving westward and made its next major stop in the enormous white pine forests that surrounded the Great Lakes. Many lumbermen went directly to the Lake region from New England. The earliest sign of their general migration from the Northeast was the purchase, in 1836, of a tract of timberland on the St. Clair River in Michigan. The buyer was Charles Merrill of Lincoln, Maine, and one after another, Eastern lumbermen followed in his footsteps. The purchase price of timberland in the Lake States was $1.25 an acre, the standard fee for land everywhere in the United States' public domain. That was 10 times what Maine timberland had cost a generation before, but well worth the price.

As in Maine, the extensive pine forests of Michigan, Wisconsin and Minnesota appeared at first to be inexhaustible. Here again, it took legions of loggers to feed the sawmills, which sprang up in Saginaw and Green Bay and smaller towns — two of them, in Michigan, named nostalgically for Augusta and Bangor. The timber barons of the Lake States soon found themselves shorthanded; they were unable to fill their logging gangs and sawmills with workers from New

England, since a great deal of timber was still being cut there. So the boss lumbermen recruited Canada's "blue-nosers," so called because of the frigid winters in their Maritime provinces. They also turned to Europe and imported "scandies" from Norway, Sweden and Finland to swell the ranks of workers.

Logging and the lumber trade boomed in the Lake States, making great fortunes for the timber barons. But long before these vast new forests reached peak production, it was perfectly obvious that here, as in Maine, the prime timberlands were bound to dwindle sooner or later. Moreover, the decline was likely to begin sooner, not later; by the 1840s, the demand for lumber was increasing just as fast as the fast-growing American population. But most lumbermen—and most Americans, too—were supremely unconcerned, supremely confident of nature's bounty. And they were absolutely right—at least by the standards of their exuberant, ambitious times. Ahead of them, far beyond the advancing frontier, lay the continent's greatest timber resource—the storied forests of the Pacific Slope and the Far West.

The ancestral forests of the Pacific Slope were first discovered not from the east, but by seaborne explorers approaching from the west. In 1579, Sir Francis Drake's globe-girdling *Golden Hind* put ashore for refitting near the site where San Francisco would rise; his crewmen chopped down trees and fashioned strong new masts and spars on which to rig their sails. About 200 years later, another English explorer, Captain James Cook, saw all too much of the towering walls of greenery that lined large sections of the coast. He sailed north searching in vain for a waterway through the American land mass to the Atlantic Ocean.

One of Cook's midshipmen, George Vancouver, returned to the Pacific Coast in 1792 as captain of the Royal Navy's *Discovery;* his expedition did much to advance the scientific investigation of the Northwest and its forests. In one of his many notable achievements, Vancouver circumnavigated the 250-mile-long island, which he immodestly gave his own name. He also explored Puget Sound, and named it (after one of his crewmen, Peter Puget) and 73 other landmarks and waterways.

Among Vancouver's crew was Archibald Menzies, a trained naturalist. During *Discovery*'s six-month cruise, he methodically explored the forests that covered Vancouver Island and the adjacent mainland shores. Menzies found many trees unknown in Europe. He collected samples — bark, seeds, twigs and pine needles — and sent the specimens back to his colleagues in London. Thus began their cooperative effort to classify the trees botanically.

In 1823, the man who was to become the greatest pioneer botanist of the Northwestern forests was dispatched from London by the Royal Horticultural Society. He was a young Scottish scientist named David Douglas, and his mission was to study the trees and send home specimens for further analysis by the society. Soon after his arrival in the Puget Sound region, Douglas made his most important find — the huge tree that was eventually to be named after him, the Douglas fir. A few botanists had already concluded that the tree was a member of the pine *(Pinus)* family, but that it was not a true pine. Douglas believed — quite correctly, as things turned out — that the tree was unique; it combined certain characteristics of the pine, the fir *(Abies)* and the hemlock *(Tsuga).* On the basis of the Scottish scientist's work, later botanists invented a new and misleading classification for it: *Pseudotsuga*

or false hemlock. By any name — and it was also widely known as the Oregon pine — the Douglas fir was unforgettable and invaluable, the most important of the 10 major species of the Western forests *(pages 24-27).* Ranging southward from British Columbia into western Oregon, and benefiting hugely from the heavy rains that fell there annually, the massive trees commonly reached a height of 250 feet, a diameter of 10 feet and an age of 700 years.

Here the Douglas fir was the dominant tree, outnumbering all its coniferous neighbors — various pines, the splendid Sitka spruce, the fragrant western redcedar, the droopy-branched western hemlock. Douglas firs were also numerous — albeit shorter and thinner — on the drier slopes of the Rocky Mountains. One factor did inhibit the long-term spread of the Douglas fir: to germinate and grow strongly, its seeds needed more sunlight than they were able to get in the tree's own towering shadow; for this reason, each dense, untouched stand of mature trees was gradually invaded by shade-tolerant hemlocks and spruces, which, after centuries, would supplant them. Nevertheless, Douglas firs were so plentiful, and grew so freely under favorable conditions, that they became the main source of the world's lumber.

The Douglas fir was a lumberman's dream. The thick trunks of many mature trees were clear, branchless wood which grew up to heights of 100 feet and more. The wood, straight-grained and tough, was ideal for heavy-duty construction of every imaginable sort and quickly won favor among homebuilders and shipbuilders alike. Used as spars, it was resilient enough to stand great stresses; fashioned into stout beams, it was long enough and strong enough to make masts for the largest sailing ships of the times. Used as tunnel stanchions, bridge girders and flooring supports, it stood up to tremendous strains. Wherever it was used, the wood of the Douglas fir held nails, screws and bolts even more securely than the sturdy oak.

David Douglas was put on the trail of another great conifer when he was shown — probably by an Indian — the cones of what was obviously a true pine. But what astoundingly large cones they were: up to two feet in length and five inches in width. Searching for the tree to match them, Douglas headed southward into Oregon country, where he found enormous forests of the species. The trees were indeed gigantic, with a height commonly in excess of 175 feet and a diameter of six feet or even greater. Their sap was as sweet as maple sugar, a characteristic which in time gave the trees their name: sugar pine.

All these facts about the sugar pine Douglas duly noted, along with proof that botanists in those days were sometimes involved in derring-do. His journal told the story:

"October 26, 1826: About an hour's walk from camp I met an Indian, who on discovering me instantly strung his bow. As I was well convinced that this conduct was prompted by fear and not by hostile intentions, I laid my gun on the ground and waved my hand for him to come to me, which he did slowly and with great caution."

To put the Indian at ease, Douglas "gave him a smoke" and a few beads. Then: "With my pencil I made a rough sketch of the cone and pine tree which I wanted to obtain. He instantly pointed to the hills 15 or 20 miles distant toward the south, and when I expressed my intention of going thither, cheerfully set out to accompany me."

At midday Douglas and his Indian guide arrived at a stand of enormous sugar pines. The big cones hung far above Douglas' reach and, since the unbranched lower trunks were too thick for him to climb, he aimed his musket skyward and shot down three cones. But "The report of my musket brought eight Indians, all of them painted with red earth. They appeared anything but friendly."

Douglas indicated to the Indians that he wanted only the pine cones, not trouble, and the Indians seemed to be satisfied. "But presently I saw one of them string his bow, and another sharpen his flint knife. To save myself by flight was impossible, so without hesitation, I stepped back about five paces, cocked my musket, drew one of my pistols out of my belt, and showed myself determined to fight for my life. Thus we stood looking at one another for perhaps 10 minutes, when one at last gave a sign that they wished for some tobacco; this I signified they should have if they fetched a quantity of cones."

With that, the Indians went off. "No sooner were they out of sight," Douglas wrote, "than I picked up my three cones and some twigs and made the quickest

possible retreat." As for his Indian guide, Douglas dismissed him "lest he should betray me."

From southwest Oregon, where Douglas found his sugar pines, commercial quantities of the species ranged southward through much of California. The sugar pine reached its optimum development about a mile high along the western slopes of the Sierra Nevada. The difficult terrain in which the species grew — granite ramparts in some areas, cruelly sharp lava beds in others — caused serious problems for the loggers who had to get the trees down from the mountains. But the sugar pines were well worth all the trouble. Their wood, light yet durable, could be put to innumerable uses. Because it swelled or shrank little when exposed to the weather, the wood made excellent shingles, doors and window sashes. Once its sweet sap had dried, the wood imparted no flavor or scent to things it touched. For this reason sugar pine became the lumber of choice for making fruit and vegetable crates and for packing tea, coffee and spices.

Still another of David Douglas' major finds was the western white pine, sometimes called the Idaho pine because 80 per cent of the stands worth harvesting were located in the Idaho Rocky Mountains. But of all the pines Douglas investigated, the most important commercially was the ponderosa, or western yellow pine, a handsome tree that often exceeded 165 feet in height and four feet in diameter. Almost everywhere in the West, from Mexico to Canada, these hardy trees grew in profusion, ranking second in lumber production only to the Douglas fir. Their strong wood, fine-grained but rough-textured, was good for almost every use; many a settler in the Northwest built his house entirely of ponderosa pine.

Douglas made several more journeys through the Northwestern forest lands, collecting and studying many other trees and plants. But unfortunately he did not live to enjoy the fame that his discoveries eventually brought him. In 1834, he set sail from California to explore Hawaii, and there, at the age of 36, he was gored to death by a wild bull.

As lumbermen later learned, the wood of other Western species was particularly well suited to specific uses. The Sitka spruce, ounce for ounce the strongest wood in the world, made the best ladders. The elegant, ginger-scented wood of the Port Orford cedar,

Totem poles carved from the tall, straight red cedars of British Columbia stand before the huts of the Haida Indians of the Queen Charlotte Islands. The Haidas also fashioned the wood into sea-going canoes.

which grew in commercial quantities only in a short stretch of the southern Oregon coast, took a fine finish and therefore was in such demand for coffins in the Orient that one naturalist facetiously suggested there was more of it underground in China than aboveground in Oregon. The western hemlock, which early loggers shunned in their efforts to get at more profitable trees, turned out to be excellent for flooring, paneling and furniture. It also became the chief source of wood pulp for the paper industry.

And then there were the redwoods, two species of them, one more unbelievable than the other. The taller of the two, standing head and shoulders above all the

forest giants except the tallest Douglas firs, was the coast redwood. It flourished in a 500-mile-long section of the fog-shrouded coastline from the Oregon border south to Monterey Bay; and after many centuries— 2,000-year-old trees were not uncommon—a good-sized specimen could tower 350 feet from a base that was 15 feet in diameter.

Its inland cousin, the Sierra redwood, was often shorter by about 50 feet but even more massive; many specimens were measured at more than 20 feet in diameter. These monsters, limited in range to 71 small groves 4,500 to 8,000 feet above sea level, thrived on the Sierra's dry summers and 10-foot winter snowfalls. Because of their sheer bulk, they were conventionally described as "the greatest living things on earth." The largest known specimen, the 3,000-year-old General Sherman tree standing in Sequoia National Park, was calculated to weigh 1,400 tons, and to contain enough wood (600,000 board feet) to build 40 five-room houses.

The coast redwoods, growing down to the very shoreline in places, had been seen from the Pacific as early as the 16th Century; Drake and other seaborne explorers could hardly have overlooked them as they sailed up the northern California coast. But the earliest description of the forest giants came from an overland traveler, Franciscan Padre Juan Crespí, who encountered them on his way from Mexico to Monterey Bay with a group of compatriots. The friar commented in the log of his journey: "The coastal plains and low hills are well forested with very high trees of a red color not known to us. They are named redwood from their color."

Crespí could not have known that the red color was due to the heavy tannin content in the trees' sap, but he did note that the wood was brittle; after his arrival on the coast, he surely saw redwood beams across the ceilings of the Franciscan missions recently established there. Those beams were presumably hewn from young, small trees or from storm-felled branches (themselves up to four feet thick). Mature coast redwoods, even those felled by natural causes, were probably beyond the capabilities of early axmen.

The inland Sierra redwoods, ignored despite the reports of a few American explorers in the early 1830s, caused a sensation when they were accidentally rediscovered in 1852. The man who thus won fleeting fame was A. T. Dowd, a hunter hired to supply meat to the gold miners at Murphy's camp in Calaveras County. One spring day in the wilderness, Dowd shot and wounded a grizzly bear. Tracking the animal to a grove of towering trees, Dowd came upon one tree so immense that he completely forgot about the bear. He hurried back to camp agog with the news, but nobody believed him. The miners chortled that old Dowd was just drunk again.

After long meditation in solitude, Dowd decided that a good lie might work where the truth had failed. Returning to camp, he announced to the men that he had just shot an enormous grizzly. This time a few of the miners decided to go along with Dowd in order to see for themselves the fearsome beast. He led them to the tree, and they stared up at it in slack-jawed amazement. "Now, boys, do you believe my story?" Dowd yelled triumphantly.

Word of Dowd's find spread. Local newspapers picked up the story and carried it east; soon scientists were traveling west to investigate. As in David Douglas' day, English botanists stole a march on their American colleagues; in 1853, they published the first scientific reports on the Sierra redwoods. Within a few years, botanists noticed the close resemblances between the two redwood species and decided that they belonged to the same genus. The coast redwood was named *Sequoia sempervirens* (always green) and the Sierra redwood was called *Sequoia gigantea* (gigantic); the name of the genus honored the Cherokee chief Sequoyah, who had devised an alphabet for his tribe. In common usage, the species growing along the coastline came to be known as redwoods, and the Sierra species as sequoias or bigtrees.

For commercial loggers, both species were a king-sized bonanza; their wood, amazingly resistant to weathering and rot, made unsurpassed shingles, house siding and railroad ties. With unflagging enthusiasm, lumbermen ravaged the Sierra groves, whose giant trees were far from numerous to begin with. The wholesale destruction was finally arrested late in the century, when Yosemite and other protected parks were established. But by that time, the sequoias had become an endangered species.

The Far West's great coniferous forests boasted many other species, including some oaks and other hardwoods. Between the Cascades and the Rockies, on ground too dry for even the ponderosas, grew the contorted juniper trees—their berries dusty blue in season, their gnarled wood good for small furniture if not much else. Among several unusual species was the 200-foot-tall western larch, one of the few "evergreens" that dropped its needles as winter came on. Odder still, and a perfect symbol of the forests' persistent life, was the skinny, ubiquitous lodgepole pine, ready-made for fence posts and telegraph poles. Dense stands of lodgepole pines led a phoenix-like existence. Many cones of the species were so tightly clenched

Visitors to California's Calaveras Grove explore the hollow trunk of a huge redwood in this 1878 watercolor entitled *Father of the*

Forest. Discovered in the early 1850s, the tree was over 300 feet from roots to crown and measured 110 feet around at its base.

shut that they would not open to release their seeds until a forest fire swept through the stand, incinerating the trees and forcing apart the cones as heat does a clamshell. Thus a new lodgepole forest would arise from the ashes of the old one.

The first loggers to harvest this family of forests were the Indians. The tribes along the blustery Northwest coast—the Haida, Kwakiutl and Tlingit, the Nitinat, Clyoquot, Cowichan and Maka—had learned from their ancestors how to fell cedars and spruces. Their implements were stone hatchets and sharpened mussel shells; by hacking and burning the base of the trees, they undermined the towering trunks and brought them crashing down. They shaped and hollowed out the trunks with scrapers made of bone and stone. Then they steamed the interior with water and hot rocks to make the wood more flexible and to widen the beam to about eight feet. Thus from single logs they fashioned high-prowed war canoes and cargo carriers up to 70 feet long. They carved the tough wood of the purplish-barked Pacific yew into durable canoe paddles and stout hunting bows.

James G. Swan, a self-educated scientist who chronicled Northwestern Indian life for the Smithsonian Institution in the 1860s, reported that the Clyoquot and Nitinat tribes on Vancouver Island were by far the best shipbuilders of the region, perhaps because they had the largest cedars growing at hand. "These canoes," Swan wrote, "are beautifully modeled, resembling in their bows our finest clipper ships." The chiefs, who had a number of slaves at their disposal, monopolized the work of felling trees to make the canoes. After the artisans had finished their work, the chiefs traded spare canoes to other tribes to the north and south for wives and more slaves. Among Indians, too, the rich got richer.

The Indians equipped their great canoes with sails of woven bark strips and ventured far out to sea to harpoon whales, which they would tow home at the end of tough lines plaited from bark or kelp strips. Because of their success at whaling and salmon fishing, the coastal tribes became wealthy by Indian standards. In turn, prosperity enabled them to have the leisure time for an immense and varied production of arts and crafts. Tribal sculptors skinned the bark off western redcedar

trees and carved totem poles with the likenesses of bear, eagle, otter and the mythical thunderbird, a winged creature that the Indians believed protected them when they were good and punished them with storms when they committed evil deeds. Brightly painted, the towering poles were erected in the villages facing seaward and served as boastful advertisements of the powers that the tribal chiefs derived from these noble creatures.

Interior tribes also did some logging, though on a more modest scale. Around Klamath Lake in southern Oregon, the Klamath Indians laboriously cut down ponderosa pines and hewed out crude canoes 15 to 25 feet long. They used the little craft for transportation, fishing and snaring waterbirds.

The sophisticated tribes of the Northwest coast were prudent loggers, cutting modestly for their modest needs, and they may well have practiced forest management long before the term was invented. Apparently the Indians realized that huckleberry bushes, which provided them with one of their basic foods, grew poorly in dense shaded forests but prospered luxuriantly in the sunlight of natural glades. So, in order to open up the sections of the forest for huckleberries and certain other semicultivated crops, they would occasionally set fire to a wooded hillside. The huckleberry bushes soon pioneered the burned-out area, whose soil was naturally fertilized by the trees' ashes. Far from damaging the forests, small local fires actually renewed them. The wind-blown seeds of mighty conifers would take root in the clearing, and in time the seedlings would grow into a strong young stand of trees—thanks to the Indian loggers.

The Northwest's first commercial logging venture was launched in 1827, six miles above Fort Vancouver on the Columbia River; in those days, England and the United States both claimed the region, which was known as the Oregon country. The fort was an English trading post, and Dr. John McLoughlin, a physician who also served as regional factor for the Hudson's Bay Company, needed wooden boxes in which to ship his trappers' furs back home to England. Also, he and the royal governor of the region, Sir George Simpson, realized that lumber would be a good source of side income if the company's ships offered it

for sale in South America and in Hawaii (then known as the Sandwich Islands).

Up to that juncture, at Fort Vancouver and the few other settlements in the Far West, logs had been cut by muscle-power alone; the gear-and-lever system that drove the saws in the water-powered Eastern mills had not yet been exported to the Pacific Coast. One log, and then another, was rolled on top of a scaffolding above a dug-out pit. While one sawyer on the scaffold straddled the log, a second sawyer stood in the sawpit below, and a long two-handled blade was drawn up and down to whipsaw the log. The man in the pit received a steady stream of sawdust in his face, and the two sawyers did well to cut as many as 150 board feet of lumber per day.

McLoughlin wanted more lumber than the whipsawyers were able to supply and, at his request, the company's home office in London sent him saw blades and gears. The doctor and his men at the trading post set up a little sawmill with a rude mechanical saw, driven by water power, that could cut 3,000 board feet of lumber per day. The logs it sawed were Douglas firs —although the trees were not yet named for David

Douglas, who had been McLoughlin's guest in his recent botanical tour of the Northwest. The doctor called the firs Oregon pine.

McLoughlin's crew of eight loggers and millhands included a few husky Kanakas—Hawaiians recruited by the company in Honolulu. Indentured for three years, they were paid the equivalent of $85 annually, plus all the salmon, sea biscuits and wild berries they could eat; at the end of their term of servitude the company offered them free return passage to the islands. The company disclaimed any responsibility for workmen maimed or killed on the job. Thus began a tradition that persisted to the end of the century: a logger worked at his own risk.

McLoughlin's mill prospered through the years; its crew increased to 28 men by 1836. However, a decade later, when England's boundary dispute with the United States was amicably settled at the 49th parallel, the treaty terms placed the sawmill site in American territory. The Hudson's Bay Company ordered its northwestern headquarters moved to Fort Victoria on English-owned Vancouver Island and the Vancouver mill was leased to the United States Army quar-

The logger's lingo deciphered

Back cut: the second and final cut made in "falling" a tree. After the initial deep undercut on one side of the trunk determined the general direction of the fall, the loggers completed the job by making a back cut on the opposite side of the tree, slightly higher than the undercut and just deep enough so that the tree toppled.

Bindle: woodsman's bed roll or pack.

Boom: a raft of logs in the water, also the line of logs or timbers strung together end to end to enclose the raft.

Bucker: lumberjack who trimmed and sawed felled trees into more manageable lengths.

Bull of the woods: the logging boss or camp foreman.

Bull whacker: driver of the ox teams that hauled logs out of the forest.

Chute: a dry trough made of wood or simply gouged in the earth, used to slide bucked timber down a hillside.

Cruise: to estimate the amount and value of the standing timber in a particular region.

Deacon seat: a bunkhouse bench—usually a split log, flat side up—that stretched in front of the bunks.

Dog: a short metal spike pointed at one end and with an eye, or ring, at the other; among many uses, a dog was driven into a log so that it could be tied to another log.

Dog hole: one of a number of small coves along California's wave-lashed Mendocino coast where the lumber schooners could put in to load cargo.

Faller: the man with whom the logging process all began—the woodsman who cut down the trees.

Flume: an inclined wooden trough carrying running water, mostly used to float sawed lumber from isolated highland mills to lower-lying mills, markets or transportation centers.

Ground-lead logging: an early method of hauling logs, by cable winched to a donkey engine, along the ground to a central "yard" or collection point. In later, more sophisticated "high lead" logging, a spar tree, still rooted to the ground, was rigged with cables and pulleys so that cut logs could be lifted at one end and hauled to the yard more easily over such obstructions as stumps and rocks.

Highballing: speeding up the logging operation. Very fast, expert lumberjacks proudly referred to themselves as "highballers."

High climber: daredevil who climbed and topped a spar tree in preparation for high-lead logging; in some operations he also served as the "high rigger" and fitted the tree with cables and pulleys.

Key log: a log that lodged in a river or stream in such a way that it caused a jam during a log drive; to clear the jam, lumberjacks set about finding and dislodging the key log.

Muzzle loader: a bunkhouse so crowded that the men of the woods had to climb into the bunks from the foot rather than from the side.

Peeler: a woodsman who stripped off the bark usually after the felled trees had been sawed into lengths.

River pigs: the men who worked on the river drives.

Skidroad: the road or path through the woods over which logs were dragged —at first by horses, mules or oxen, later by mechanical means; also came to apply to the saloon-packed district in lumber towns where crews went to carouse in their free time.

Snipe: to shape or bevel the front end of a log with an ax, making it easier to haul the log along a skidroad, or slide it down a chute.

Splash dam: constructed across a lazy or shallow water course to build up a head strong enough when released to wash logs downstream.

Springboard: a sturdy plank notched into a tree to provide a working platform for the faller; thus elevated, early fallers avoided the swollen, sometimes pitch-saturated base of the tree.

Steam donkey: a portable steam engine equipped with cables and one or more revolving drums that found a score of uses in the industry, particularly in yanking felled logs through the woods to a central collecting yard.

Timber beast: wry nickname for anyone who worked in the woods.

Tin pants: a timber beast's waterproofed, heavy canvas pants.

Turn: any unit of logs, from one to half a dozen or more, hauled out of the woods, either by animal power or by donkey engines.

Widow maker: a treetop, heavy limb or chunk of bark dangling loose and ready to fall on an unwary woodsman.

termaster for $14,000 a year. By the mid-1840s, the straight, strong lumber of the Northwestern forests was well known and much in demand throughout the Pacific basin, from South America to Australia and even China. And increasing numbers of Americans were trickling west to participate in logging operations. In 1843, when the first sizable group of American farmers (about 1,000) entered Oregon, a handful of Middle Westerners put a mill into operation on the Columbia River opposite Puget Island. That same year a sawmill owner named Henry Hunt, with all his machinery in ox-drawn wagons, arrived from Indiana after an arduous trek across plains and mountains. Hunt built a new mill at Cathlamet Point on the Columbia River and opened for business in 1846.

By now California was also enjoying the first fruits of a budding lumber industry. Hand lumbering had begun commercially in that timber-rich Mexican province as early as the 1820s. California's first professional loggers were discontented seamen from the foreign ships that put in at Monterey Bay and San Francisco (then called Yerba Buena). The seamen, violating their contract to serve for the whole voyage, jumped ship and headed for the pine and redwood forests in the coastal hills, there to hide out until their ships had left the California coast. The fugitives, who were subject to imprisonment if they were caught, were desperate men and, as a California resident of that time observed, "No person dared to go after them, for if they did, they would never return alive."

To support themselves, the ship jumpers set themselves up as whipsawyers and shingle splitters, selling their output to any merchant or shipper who would handle it. One of the first seamen-turned-loggers was an American named William Smith, later and better known as Bill the Sawyer. Smith jumped ship at San Francisco around 1818, and went to work nearby, whipsawing among the San Mateo redwoods. There he was joined two years later by an English mutineer, and in 1821 by a demoted English naval officer. Similar groups of fugitives coalesced into California's first foreign colonies. A motley collection that sprang up near the Santa Cruz Mission grew into a village of 150 people by 1835.

The fugitive enclaves were organized and given a sense of direction by the arrival in 1833 of an American enterpriser by the name of Thomas Larkin, who operated out of Monterey. Larkin paid the whipsawyers about $40 per 1,000 board feet of redwood lumber, and began shipping cargoes on his bark *Don Quixote* to merchants in Los Angeles and Santa Barbara. In 1835, Larkin's half brother, J. B. R. Cooper, built California's first sawmill in what was to become Sonoma County; he combined with Larkin in marketing the mill's lumber output. Meanwhile Larkin's export business expanded and prospered. By the early 1840s he had overshadowed his local competitors and was shipping a large variety of merchandise, including lumber, to Hawaii, to Mexico and to the Pacific coast of South America.

Prospects for West Coast logging, and for the West Coast itself, improved still further in the mid-1840s. In 1845, California residents put an end to the backward Mexican rule, driving out the last governor. Then the following year, the United States went to war with Mexico; in July American naval forces captured Monterey, the capital of California, and claimed the whole territory for the United States. In 1848 Mexico finally ceded California to the United States. With the Canadian border dispute finally settled and the English acquiescent, the aggressive Americans were free to expand, economically as well as territorially, over the whole Pacific Slope.

However, West Coast logging was still a small-scale export trade, and so it would remain until a vital missing ingredient was added: a substantial local market for lumber. In 1847, the entire West Coast population—including the English settlements north of the border, the American logging communities and the American farmers in Oregon, the Mexican towns and foreign enclaves dotting the California coast—amounted to no more than 25,000 people. Behind them, there was nothing but wilderness for some 2,000 miles, all the way east to Missouri.

In the normal course of events, it might have taken a half century or more to build up a substantial population in the Far West. But in the winter of 1848, something distinctly abnormal occurred—an epoch-making discovery that triggered the biggest and swiftest mass migration in history. And it all began, appropriately enough, at a logger's mill on the southern fork of the American River in California's Sierra foothills.

Lilliputian visitors to the redwoods' realm

From their discovery in 1852, California's redwoods fascinated Americans. So many flocked to the Sierra Nevada to glimpse them that within a year the first resort went up; others followed, many located within sight and sound of logging operations.

The loggers spared some of the largest redwoods, which were given names such as Hercules or George Washington. "The height of enjoyment," wrote a visitor in 1870, "is to lie down on your back under the full moon and look up, say ten feet at a look, till the eye has travelled all the way up to the top —over 300 feet." Others, loggers and visitors alike, picnicked and held dances on the great stumps, explored rooms hollowed from the cavernous butts and rode carriages through the tunnels in the trunks.

No one who ever stood next to a redwood was likely to forget the experience. But photographers came to the aid of memory's eye, capturing on film Lilliputian man's encounter with the largest living things on earth.

Some 1890s visitors pause for a commemorative photograph of their coach ride on the famed Fallen Monarch of Mariposa County.

41

In a Tulare County logging camp, square dancers prepare to heel-and-toe on a redwood-stump floor.

Two loggers on horseback emphasize the enormous undercut of an about-to-be-felled Sierra redwood.

A coast redwood stump provides a setting for a grammar school class picture in the 1890s.

2 | The great timber rush

In the early 1840s lumbermen had scarcely penetrated the forests of the Pacific Coast. A few seaside sawmills scattered from California to Washington provided lumber for local settlements and for a small export trade. But they were crude affairs, some consisting of little more than a pit and a platform where a two-man team could work an eight-foot-long whipsaw.

Then, in 1848, came the event that forged an industry almost overnight. Gold was discovered in California, and the rush to mine it triggered the greatest mass migration in United States history. Suddenly, San Francisco and other boomtowns needed lumber far beyond the local woodsmen's capacities. At first, the wood came from far away, mostly from Maine by ship. But the irony of bringing lumber halfway around the globe to a land carpeted with the mightiest trees on earth was not lost on men of vision. Soon a new breed of American entrepreneurs rose up to build empires of wood from the forests of the Far West.

The loggers staked out vast tracts along the northern California coast and on up into Oregon and Washington's Puget Sound. The early mills were joined by more sophisticated installations, with well-organized teams to fell the trees, beasts of burden to move the logs, and steam-powered circular saws that could buzz through many thousands of board feet a day.

But no matter how much timber the woodsmen cut — and before long production soon was measured in many millions of feet annually — they hardly seemed able to make an impression on the forests. Trees grew in such profusion that a traveler at the time could only exclaim: "Timber! Timber! Till you can't sleep!" For decades the lumberjacks thought it would last forever.

At a mill site during the late 1880s, a steam-powered circular saw cuts shingles from redwood

shake bolts. A bull puncher *(foreground)* has just led in his team with more logs. The fire at lower right is a sawdust pile being burned.

The opportunists who carved their fortunes from wood

John Sutter had already earned a reputation as one of California's more enterprising businessmen when he reached a fateful decision in 1847 to erect a sawmill and make his mark in lumber. An ambitious German-born Swiss who had immigrated to California in 1839, Sutter quickly built an impressive pastoral empire on the 50,000 acres of land granted to him by Mexican authorities in the fertile Sacramento Valley. His headquarters were located at the confluence of the Sacramento and American rivers in a stockaded village he named New Helvetia, but which most everyone else called Sutter's Fort. There, at the age of 44, he ruled a domain that included a thriving trading post, hundreds of acres of wheat fields, and vast herds of cattle, sheep, hogs, horses and mules.

Sutter had his flaws as a businessman. In his race to grow, he frequently found himself overextended and deeply in debt. Nevertheless, his long-range prospects seemed bright indeed. In the last year or so a growing stream of immigrants had begun to descend on the fort, which marked the end of the long California Trail across the mountains and deserts from Missouri. At Sutter's, the weary pioneers rested and replenished their supplies before fanning out in search of suitable farmland and homesites. The industrious Sutter planned to expand his trading post and build a large new gristmill to grind his wheat into flour for sale to the settlers. But he found this and other plans stymied by an irritating problem: the chronic shortage of sawed lumber.

Though the nearby foothills of the Sierra Nevada were blanketed with pines and cedars, there were no sawmills in the vicinity that could convert the trees into useful planks. The closest mills were 75 to 100 miles away, clustered around San Francisco Bay, and the task of securing lumber from them was both costly and time-consuming. Periodically, Sutter would have to load a boat with trading goods — hides, tallow, furs and wheat — and send it down the Sacramento River to the bay area to barter for sawed lumber. The round trip took a month and Sutter was often unhappy with the quality of the boards he received. Worse yet, the cost was, he felt, exorbitant.

At one point in 1841, desperate for lumber for his own uses and to sell to settlers, Sutter went so far as to buy the Russian colony at Fort Ross, north of San Francisco Bay, for $30,000. He dismantled the buildings and shipped the redwood planks to his fort. But they lasted only a short while and Sutter still faced the usual problem of heavy demand but scant supplies.

It was against this background that Sutter, in August of 1847, entered into partnership with a 36-year-old millwright named James Marshall, a recent arrival from New Jersey by way of Oregon. Their agreement was that Marshall would find a site and build a sawmill from which a large, dependable supply of high quality lumber could be rafted downriver to the fort. Marshall would run the mill in return for half of the lumber cut. The partners shook hands and a few weeks later Marshall reported that he had found just the site. It was on the south fork of the American River, about 50 miles northeast of the fort; the hills all around were thick with timber. Late in 1847, Marshall began constructing the mill, which was to be powered by water diverted from the river by a canal-like millrace. Back at the fort, Sutter eagerly made plans to receive the first lumber from his new mill in the spring.

But he was destined never to see any profit from his precious wood. On the morning of January 24, 1848,

Lumber and the ships to haul it dominate the San Francisco waterfront in 1865. More than 100 million board feet of lumber from mills all over the West Coast were sold on the city's booming market that year.

Marshall was inspecting the mill-race when something caught his eye — a glint, a reflection of sunlight in the water. He bent over and idly picked up a few bits of yellow substance resting on the bottom. Gold! The shiny pebbles were pure gold, the first tiny nuggets of a fantastic strike that would transform Sutter's simple little sawmill into the epicenter of the greatest gold rush the world had ever seen.

It proved to be the ruination of John Sutter. His farm hands and laborers threw down their tools and dashed to the millsite to pan gold. As waves of frenzied gold seekers washed over the Sacramento Valley, the prospectors brazenly invaded Sutter's lands at the fort, trampled flat his wheat fields and drove off his untended livestock. Sutter himself tried panning for gold, did poorly and increasingly took to drink. Before long, his never-patient creditors began to close in on him. At last, in the summer of 1850, John Sutter sold his remaining interest in the fort for $40,000 and retreated to his small farm on the Feather River to the north. Eventually, he left California altogether and died in Washington, D.C., at age 77, poor, embittered and still trying to collect a claim for $125,000 in damages from the federal government, which he argued had profited hugely from the gold rush. "It has turned out a folly," he once wrote, explaining that, but for the gold, "I would have become the wealthiest man on the Pacific shore."

There was an irony of heroic dimension in the tragedy of John Sutter. His yearning for a lumber business set in motion the cataclysmic events that destroyed him. Yet it was the gold at Sutter's Mill that set the stage for an even mightier and more fateful national drama: the growth of a lumber industry that surpassed anyone's wildest imaginings. Gold triggered the mass migration of Americans to the Far West, populating the empty land and opening an insatiable market for the vast timber resources at every hand.

In a decade, from 1850 to 1860, California's population soared from 93,000 to 379,000 and

John Sutter

the people had to have lumber — for homes, for stores, for industries. San Francisco was at the hub of it all and its population rocketed from 2,000 at the beginning of 1849 to 55,000 by 1855. Six times, in the space of 18 months between December 1849 and June 1851, the ramshackle, jerry-built boomtown burned to the ground. And each time, it was frantically rebuilt, in the process consuming millions upon millions of board feet of lumber.

In 1849 there were only 10 sawmills operating in all of California and they could cut no more than five million board feet each year — hardly enough to make a dent in the sudden demand. Pine and redwood that had been selling for $40 to $50 per thousand board feet in 1848 shot up tenfold in a single year. For a brief period, San Franciscans were scrambling to buy ordinary planks at the astronomical price of $1 for a single board foot.

So precious was wood that miners hoarded it to build their sluices, troughs and other equipment — often at the expense of their homes. The mining town of Stockton contained only two wooden houses in 1849, and a resident explained, "Everyone here lives in tents, as lumber is too expensive and ordinarily no boards are available." Sutter's ill-fated mill met its end, when prospectors picked it apart, plank by plank, for their mining needs.

Canny New England ship captains, sailing around Cape Horn to California carrying gold seekers, filled their holds and piled their decks high with precious lumber, which they sold at huge profits. Between 1849 and 1852, no fewer than 90 vessels made the six-month voyage to California from Maine alone, and all of them carried lumber in one form or another. The barkentine *Suliot,* arriving in San Francisco early in 1849, was fitted with passenger berths made of hemlock that had cost $10 a thousand feet in Maine; ripped out, the boards sold for $300 a thousand feet.

Lumber was also arriving in San Francisco from the farthest reaches of the globe — from Norway, Australia,

Sutter's humble sawmill at Coloma on the American River, where gold was first discovered in 1848, was an abandoned shell when this scene was painted a few years later at the height of the great gold rush.

Chile. But anyone with half an eye could hardly fail to grasp what a grossly inefficient trade it was. With billions upon billions of board feet of the finest timber carpeting the hills up and down virtually the entire West Coast, it was obvious that the truly large profits lay in utilizing local resources. A shrewd and enterprising man with a sawmill and the vessels to transport lumber to market could make his fortune many times over.

Heretofore West Coast logging had been a rudimentary, seasonal business. Lumberjacks and millowners worked only during the warm, relatively dry six months from spring through fall. Everyone halted operations when the chill winter rains arrived and made it difficult to haul the huge trees out of the sodden forests. But with soaring demands for wood, mill operators everywhere sought to build up large stock piles to keep the saws buzzing 12 hours a day, six days a week the entire year round. (With increasing mechanization in the forests, logging continued around the calendar as well.) And large fleets of specially designed schooners plied the coastal shipping lanes in fair and foul, transporting the lumber to market.

Ended, too, was the hit-or-miss method of securing logs for the mills. In the old days, sawmill operators bought most of their timber from pioneers who felled the trees while clearing land for farms, or who engaged in logging as a sideline. But now teams of professional woodsmen invaded the forests, and sharp-eyed "timber cruisers," or "land lookers," prowled the Pacific Coast, seeking out the most desirable stands to cut.

Some companies bought the land outright; others thriftily paid only for "stumpage"—the right to clear the timber. And there were any number of other ways in which lumbermen managed to lay their hands on supplies. One favorite device was to pay mill workers and the crews of lumber schooners a few dollars to file homestead claims of up to 160 acres per man on choice land, which they promptly turned over to their bosses. Another popular tactic after the Civil War was to buy up the federal scrip—redeemable in government-owned land—that had been issued as a bonus to United States Army veterans. A few lumbermen—the unscrupulous ones—dispensed with the niceties altogether; they simply marched in and committed trespass: logging private and government land without so much as a by-your-leave.

However the loggers managed to secure their trees, by 1859 the Pacific Coast as a whole was able to boast a production of more than 300 million board feet of lumber annually compared to a mere 25 million a decade earlier. It was, however, too much to expect the industry to rise in a single unbroken upward thrust; from time to time, as production expanded in quantum jumps, lumber companies suffered setbacks from glutted yards and falling prices. But these dips were only brief interruptions in the amazing growth of the industry. By 1880, lumbermen were cutting and selling close to 700 million board feet annually.

It took the talents of a fascinating crew of pioneering barons to build this great industry. In common, they were bold, powerfully willed, ruggedly individualistic men. None ever gained dominion over the others; the raw material that they dealt with and the markets that they supplied were too gigantic and widespread to be graspable by any monopoly.

In the winter of 1849 the 140-ton brig *Oriental,* out of East Machias, Maine, spent 171 days sailing down the Atlantic, around Cape Horn and up the Pacific to San Francisco Bay. All of the 11 passengers were bound for California gold fields and the chance at sudden fortune. Her captain, William Chaloner Talbot, had come to stay also—but not as an Argonaut.

Lumber was Talbot's business, as timber and sawmilling and seafaring had been the way of life in his family for generations. Among his cargo on *Oriental,* he included house frames and shingles, heavy timbers, plus 60,000 feet of sawed lumber—all for sale in California. His voyage was a step in a long-range venture, far bigger and far different than he had any reason to suspect at the time. He had sailed West to trade; he would stay to help establish and run the firm of Pope & Talbot, one of the earliest and most durable of logging, milling and shipping dynasties to be found on the Pacific Coast.

At 33, William Talbot was already a seasoned captain and lumber trader, whose ships had carried good Maine pine to Europe and Central America. His had

been a prosperous and satisfying life in East Machias. He had a fine wife and two pretty daughters. He shared ownership and management of the family dock and lumber mill with his father and two brothers. Nevertheless, he had listened with rising interest to the captains returning from California, with their accounts of fabulous prices paid for lumber. It did not tax his Yankee business sense very hard for him to perceive an opportunity of great promise in the West.

Nor was he the only Talbot with such ideas. No sooner had Captain William cast off the lines of *Oriental,* on September 15, 1849, and sailed south out of East Machias, than his brother Frederic set his own adventurous plan in motion. Three years younger at 30, Frederic had often talked about California with his lifelong chum Andrew Pope, a bright and industrious man of 29. Pope had settled in Boston, helping to manage lumber sales, ship charters and finances for the family firm of S. W. Pope & Company. When Frederic Talbot came to him with a plan to go West, Pope quickly agreed to join the venture. With another friend, J. P. Keller, a ship captain from East Machias, they proceeded to New York to start their voyage.

On October 16 they boarded the steamship *Ohio* on the first leg of a journey to California via the narrow Isthmus of Panama. By traveling overland across the Panama jungle and boarding the little steamer *Oregon* on the Pacific side of the Isthmus, they were able to cut months off the time it would have taken to sail clear around South America. In fact, when *Oregon* docked in San Francisco on December 1, 1849, Captain William Talbot's *Oriental* was barely halfway through its long journey.

To the three men from Maine, accustomed to tidy and conservative towns, San Francisco seemed to be a scandalous mess. But it was just as obviously a city with a fantastic future. Thousands of would-be miners streamed through it on their way to the gold-flecked Sierra foothills to the northeast. Thousands streamed back when bad weather or bad luck frustrated their search—or when their pouches were filled with nuggets and gold dust. With all their comings and goings, the city's population ebbed and flowed like the tide.

Within a week Pope and Talbot were in business, with Keller as a partner. For an outlay of $500, they bought an old longboat and started a lighterage operation, carrying ashore cargo from sailing ships unable to find docking space. Business was so good that they quickly added two large scows and a yawl, as well as a small sloop that Captain Keller used to carry freight between San Francisco and nearby Stockton.

But the lumber business was always the main chance to which they had committed themselves. For $250 a month, the firm rented a small beach lot and opened a yard stocked with wood purchased from ships in the harbor and brought ashore on rafts. Then, on the morning of March 2, 1850, the brig *Oriental* sailed into San Francisco harbor with all passengers safe and sound —and of equal importance to her captain, not a stick of lumber lost en route.

The very last people on earth Captain William Talbot expected to see in San Francisco were his brother Frederic and his friends Pope and Keller. But when the shock had worn off, William readily agreed to join the Pope & Talbot partnership. It seemed far more promising than the arduous, once-a-year run from Maine to California. Besides, with plenty of other ships willing to risk the long voyage around the Horn, it seemed more sensible to concentrate on selling lumber rather than simply delivering it. The lumber market in San Francisco was experiencing one of its temporary lulls, so Frederic Talbot volunteered to reload the Maine pine onto a river steamer and take it to Sacramento. There, a short distance from the fort where John Sutter had vainly tried to establish a lumberyard, Talbot sold the lumber for around $100 per thousand board feet—probably half of it clear profit.

In the heady excitement of the time it seemed that everywhere the partners looked or went there was money to be made. With the proceeds from *Oriental's* cargo, they ordered more lumber from Eastern suppliers. They also sent *Oriental*, with Captains Keller and William Talbot taking turns in command, on short trips to local sawmills. But clearly, what they really needed was a setup where they could cut and sell logs of their own. They agreed that it had to be on tidewater, where the mill's products could be easily loaded directly onto ships belonging to their firm. The question was, where?

California did not appeal to them. People who had explored the redwood coast to the north reported that the only safe haven there for ships was Humboldt Bay, where a constantly shifting sand bar lay athwart the entrance, making every passage a gamble. There was plenty of timber along the Columbia River, hundreds of miles farther north, but its mouth likewise was made hazardous by surf and sand bars.

They looked to Puget Sound, the northern inland sea where less than 100 white people had settled. Its gateway was the stately Strait of Juan de Fuca, the middle of which was the newly drawn boundary between the United States and Canada. But the sound and its 2,000-mile protected shoreline, inside the broad strait, were still so little known in California that a San Francisco vessel, *G. W. Kendall,* actually had been sent up there in 1850 with the assignment of chipping a cargo of ice from icebergs; her owners—unaware of the warming Japanese current—had reasoned, not illogically, that since the waters were in the same high, cold latitude as Newfoundland, there had to be great islands of glacial ice. The sound proved to be ice-free the year around, of course, and so *Kendall* had returned south with a load of pilings, cut at the water's edge by pioneers who were clearing trees at homesites.

The profitable voyage of *Kendall* and other small vessels convinced the partners. Though they had yet to see the place for themselves, they made up their minds: San Francisco would be their home port and headquarters, and the Puget Sound area would be their source of supply.

Preparing for their commercial invasion, the partners returned East for about a year and a half to wind up their affairs in Maine. Frederic Talbot, evidently homesick for New England, had already returned East in December 1850 and he decided to remain there to help manage the family interests. His brother William, with Andrew Pope and Captain Keller, arranged to take their own families West permanently. In the spring of 1853, while Keller was coming around the Horn with his family, mill hands and machinery aboard *L. P. Foster,* Pope and Talbot returned to San Francisco. Pope took over operation of the lumberyard, and William Talbot sailed north to explore the sound in *Julius Pringle,* another vessel they had acquired. Before Keller arrived, Talbot wanted to find the ideal site to put up a sawmill for their newly formed Puget Mill Company.

Leaving *Pringle* at anchor in Discovery Bay, about 70 miles inland from the rocks at Cape Flattery that

William Talbot and Andrew Pope were heralded as the "lumber kings of the Pacific Coast" by a San Francisco newspaper in 1875, just 22 years after erecting their first mill on Puget Sound. Sharing unheralded in the huge profits was Cyrus Walker, who had become a partner in 1863.

CYRUS WALKER

WILLIAM TALBOT

ANDREW POPE

guarded the Strait of Juan de Fuca, Talbot cruised hundreds of miles along the shoreline in a little sailboat he had carried on deck. Meanwhile, in a borrowed Indian canoe, a new recruit to the enterprise paddled up and down myriad forested inlets. The man was Cyrus Walker, who on the spur of the moment had joined the enterprise back in New York and would end up spending the rest of his life with it.

Walker, a schoolteacher, was a 26-year-old native of Madison, Maine, whose goal in life was somehow to amass the sum of $50,000. Certain that he would never achieve his ambition as an educator, he had gone to New York in 1852 and booked passage to seek his fortune in Australia. He changed his mind when he saw the wretched condition of the ship, sold his ticket and was sitting glumly in a hotel lobby when he struck up an acquaintance with one of the millwrights who had been hired by Talbot and was preparing to join him. Having nothing better to do, Walker sailed for California, where he and Talbot immediately hit it off and went exploring together.

Talbot and Walker found their millsite on the east bank of Hood Canal, a natural channel that George Vancouver had discovered in 1792. In sheltered Gamble Bay, the water was deep enough for ocean-going ships and there was a flat, sandy spit of land big enough for the mill buildings. "Teekalet," for brightness of the noonday sun, was the Indians' name for the place. *Pringle* was brought to the site and unloaded. A crew of 10 was left ashore to build a bunkhouse, a cookhouse and store, and to prepare the foundation and framework for the Puget Mill Company's first sawmill. Then Talbot took his ship over to Seattle, a townsite less than a year old on the east shore of the sound, and loaded her with lumber and piling from the brand-new steam sawmill owned by Henry L. Yesler. It was the first of thousands of shiploads that Pope and Talbot would take from a land whose trees would long outlast California's gold—and in fact would long outlast all the early generations of timber barons.

On the way out of the sound into the Strait of Juan de Fuca, *Pringle* chanced to meet *L. P. Foster,* then 154 days out of Boston. As the two sailing ships stood bow to bow, Captain Talbot shouted instructions to Captain Keller about where he should deposit the engine, boilers, saws and other machinery for the mill. They said goodbye and Keller proceeded on to Teekalet, where his wife and daughter were the first white women ever to step ashore.

As of that fall of 1853, Pope, Talbot, Keller and their Puget Mill Company were in business, using a steam-driven saw to cut a modest 2,000 feet of wood a day from logs felled on the spot or bought from settlers. Within a few months production doubled and then redoubled, and before long 38 saws were cutting logs. A second mill was put into operation. By 1857, they were cutting eight million feet a year—enough to fill a ship each week. After one visit to the sound from the office in San Francisco, Andrew Pope wrote his Eastern relatives, "It's the greatest place to make lumber I have ever seen. Mills running night & day & lumber goes off just as fast as we can make it."

As production increased, Pope and Talbot aggressively sought out foreign markets. Their ships and crews were equally at home in Valparaiso, Honolulu, Manila, Hong Kong, Shanghai and Sydney. With Pope minding the business in San Francisco, Talbot often took the bridge of a company schooner. Though the firm developed a network of overseas agents, it also shipped venture cargoes—unordered loads for sale to dealers in cities where the company had no agents. If no bargain could be struck, the Pope & Talbot captain would rent dock space, divide the cargo and auction it off in small lots.

Upon the death of Captain Keller in 1862, Cyrus Walker became mill manager at Port Gamble, as Teekalet was renamed. And finally, after a decade of hard work, he was offered a junior partnership with a 1/10 interest in the Puget Mill Company, which he purchased for the carefully calculated sum of $30,044.74.

By now the company owned close to 35,000 acres of timberland and had bought stumpage rights to thousands more. Ever more efficient saws were producing 18 million feet of lumber annually, a staggering rate for the day. And Pope & Talbot boasted a fleet of 10 vessels to take it all to market. From that solid base, all through the 1860s and 1870s, the company continued to grow under the tenacious, conservative direction of the two partners. Both Pope and Talbot shared a horror of debt, ran their enterprises on a cash basis, and slowly built their holdings of timberlands, their fleet and their sales volume as well. They took careful fliers—in cattle, gold stocks, real estate—only when they had money to spare.

At the end of their first decade in the city, they had built adjacent homes on Folsom Street, in the then fashionable South Park district; Talbot's was the first house in town to have running water. In the 1870s, when their mills were shipping an incredible 40 to 50 million feet annually, spacious new Pope and Talbot mansions went up in the more desirable Van Ness neighborhood. Both men died millionaires—Pope in 1878 and Talbot in 1881—and the solid organization they passed on would prosper for many years.

The heart of the business was—and would continue to be—the milling complex at Port Gamble on Puget Sound. Under Cyrus Walker it had become the very replica of a New England coastal village. Many of its people were Congregationalists from Maine, and its house of worship was a faithful copy of their church in East Machias. Its unmarried mill hands and loggers lived in a large bunkhouse that housed up to 400 men; the married ones lived in neat, company-built white frame houses. By the 1880s, all the structures were equipped with running water that flowed from a spring in the hills. Something else was remarkable about Port Gamble: the trees shading its streets were not native to the West Coast; they were spreading maples whose seedlings Mrs. J. P. Keller had lovingly imported from Maine to add one more touch of New England.

Across Gamble Bay, the town had a suburb called Little Boston, where the company housed its Indian workers. "It was a common sight," recalled one observer, "to see a string of rowboats making their way across Gamble Bay, filled with sailors intent on appraising the charms of the squaws of Little Boston." Somewhat against the principles of the teetotaling but tolerant Walker, the company store sold liquor. What bothered him more was the town's independently owned Puget Hotel, where loggers, mill hands and visiting sailors could find card games in the saloon bar and loose women upstairs. Men gambled there the whole weekend until the Monday morning whistle summoned them to work. Walker solved that problem by having Pope and Talbot's firm buy the hotel and put it in the tight control of a company-paid manager.

In those cash-short times, many lumber companies paid their men in scrip, or vouchers good at the company store. The Puget Mill Company prided itself on its liquidity—a man could get his pay in cash every day if he wanted, or he could let it lay. When one old-timer finally decided to withdraw his accumulated wages, it came to $8,000 in silver 50-cent pieces. He piled the money into a wheelbarrow and took his hoard to the bank in Seattle on a company ship.

Port Gamble and nearby Port Ludlow, a companion company town, were home ports to a fleet of Pope & Talbot tugboats. Besides towing rafts of logs to the mills, the steam-powered tugs were profitably used to tow windjammers into and out of the strait and around the sound ports, where they could not always rely on wind power. In both roles, the tugboats were essential to the business of sawing and shipping lumber.

Any tugboat-launching was a big event on Puget Sound and July 22, 1884, the day the new *Tyee* (Chinook for chief) hit the water at Port Ludlow, was especially eventful. She was something different, a coal burner where most local tugs burned wood; coal took less space and there was a plentiful supply, from new mines around the sound and at Nanaimo on Vancouver Island. The most powerful tug yet built in America, *Tyee* was 140 feet long, with a hull two feet thick, and Cyrus Walker had personally selected the timbers for her beams and planking.

The company mills were closed for *Tyee*'s launching, and people came from all over the sound as guests of Cyrus Walker and the company. There was a gargantuan feast of roasted beef sides and hams, caldrons of baked beans, bread and pies and cakes and barrels of beer. There were footraces and a baseball game pitting the Port Gamble and the Port Ludlow teams, though there is no record of who won.

Late in the day men knocked the blocks from under *Tyee,* a child broke a bottle against the bow, and the christened craft slid down the shipways stern first. A little tug was standing by to bring *Tyee* back to the yard. But before it could get a line aboard, some pranksters from Seattle managed to do so and, in tow of their small workboat, *Tyee* started moving out into the open sound. A comic-opera chase ensued, Cyrus Walker leading it from the deck of the tug *Yakima*. When the intruders' towline parted abruptly and their boat tipped wildly in the water, he helped fish out three of the hijackers who had been thrown into the

bay. With no hard feelings, he brought them ashore to dry out, and eat and drink some more. The boss was an upright, but not a censorious man.

As befitted the head man at the mills, the finest house in Port Gamble was Walker's. But it was a bachelor home until, on June 30, 1885, at the age of 57, he brought a suitable bride back from San Francisco; she was William Talbot's daughter, Emily, 20 years Walker's junior. "We arrived home last night at 10 o'clock," he wrote to her family, "and notwithstanding the late hour we were received with the blowing of whistles, ringing of bells and firing of guns."

The house burned down that year and he decided to build anew at Port Ludlow. The Walkers opened the home they christened Admiralty Hall in 1887.

Standing within earshot of high-pitched mill saws and within view of Ludlow Bay, the new mansion was the equal of any townhouse in San Francisco. The great front doors slid open and shut like those on a ship. The woodwork was cut at the sawmill from the finest fir, and most of the furniture was handmade from black walnut shipped from Maine. Out on the front lawn there were a spacious croquet ground and a brass cannon, a relic of the War of 1812. The piece was fired at sunrise every Fourth of July, and it boomed at any time of any day to welcome a lumber vessel putting in at the mill wharf.

Cyrus Walker was typical of early-day boss loggers and sawmill operators, who could themselves do anything they ordered someone else to do—from felling a

59

The smoking chimneys of two sawmills, busily ingesting a boom of logs, dominate this view of Teekalet—later renamed Port Gamble—in the 1860s. By then, Pope and Talbot's New England-style mill town on Puget Sound was shipping around 19 million board feet of lumber annually.

61

tree to jockeying a tugboat. He could, and did, climb a mill chimney cracked by an earthquake and fasten strap-iron belts to secure it. A visitor from Seattle once found him sitting on a pile of old bricks, helping his workmen chip off the mortar. He habitually bent down to pick up nails as he made his rounds, and to lessen the chance of fire he insisted that sawdust flying from the mill saws be swept up; no sawmill in his charge ever burned. That was a wonder: for years, until electric lights came along, the dark interiors of the mills were lighted by flickering, smoking "tea kettle" lamps that burned smelly fish oil.

After 12-hour days in the mills, Walker spent many evening hours on his accounts and correspondence be-cause he did not feel right unless he had worked enough to be physically tired at the end of the day. He kept it up until his death in 1913 at the age of 86. By then, the outfit he had joined 60 years before was shipping almost 100 million feet of lumber a year all over the world — including stout planks of fir to his native Maine.

In 1852, the year before William Talbot had sailed for the first time into the lovely bay at Teekalet, a heavy-set man of 41 beached his small dugout canoe at Elliott Bay on the eastern shore of Puget Sound. He walked up to high-tide level, where some men were rolling logs into the water, and introduced himself as Henry L. Yesler, a millwright from Baltimore, lately

Champions of the "Sawdust League," Port Gamble mill hands met rival teams on Sundays. A winning pitcher could find a job at any mill.

of Massillon, Ohio. He explained that he was looking for a place to build a steam sawmill.

The loggers replied that they were sorry — claims already had been staked to all the waterfront in the Elliott Bay neighborhood. Yesler thanked them politely and started back to his canoe. Watching him go, two men leaned on their axes and held a hasty conference. They were David S. Maynard and Carson Boren, and the land on which they stood was already the site of the infant town of Seattle.

It was a lonesome place, with just a handful of houses and a general store. Once in a while a boat stopped to take on spars and pilings cut by the settlers. But that was about the extent of the lumber business around Seattle; there was no steam sawmill on the sound.

It took Maynard and Boren only a few minutes to conclude after all that a steam sawmill would not be a bad local asset. They halloed to Yesler, calling him back. He remained for 40 years, and the series of sawmills he built were the foundation on which a great city grew. He turned out to be a complex and contradictory man, overgenerous at times, cantankerous and quarrelsome at others, a chronic borrower and a poor businessman compared to Talbot and Pope. And yet he wound up being hailed as the Father of Seattle, the metropolis of the Northwest.

Puget Sound was still part of Oregon Territory in 1852 and, under the Oregon Donation Land Law, a single man could lay claim to a quarter section, or 160 acres, of free land. As a married man Yesler could claim twice that much, even though his wife and child were still back in Ohio. Carson Boren and another settler generously adjusted their claims so that Yesler could stake out a 320-acre parcel, shaped like an L, running from the waterfront up into the heavily timbered hills overlooking the sound. When Yesler left for San Francisco to pick up his mill machinery, a group of obliging neighbors pitched in to build an open shed to house his boiler and saw, and a large, sturdy log building to serve as a cookhouse. News of the new sawmill quickly got around, and the Olympia *Columbian,* the only newspaper in the region, was exultant:

"Huzza for Seattle! It would be folly to suppose that the mill will not prove as good as a gold mine to Mr. Yesler, besides tending greatly to improve the fine townsite of Seattle and the fertile country around it, by attracting thither the farmer, the laborer, and the capitalist. On with improvements!"

Copies of the paper reached Seattle, which had no roads going anywhere, via a waterborne delivery service called Moxlie's Weekly Canoe Express.

By the spring of 1853 the mill's circular saw was slicing wood at the rate of nearly 10,000 feet a day. At first, logs were floated over to the mill by settlers who were opening small clearings on their claims around Elliott Bay, and much of the wood went back to them in the form of lumber for their houses and barns. By summer Seattle could boast of some 20 wood-frame houses. Part of the mill's early production went into a small hotel, the Felker House, built by a sea captain confident of the city's future and run by an innkeeper named Mary Ann Conklin. Wife of a whaling captain and a veteran of other waterfronts, Mary Ann was a large, powerful woman who kept order in the Felker House with torrents of profanity and barrages of well-aimed kindling wood. Perhaps that was why she was also known as Madame Damnable.

To secure a bigger and steadier supply of logs for the mill, Yesler cleared a strip of land on his claim going straight up the steep slope of what came to be known as Seattle's First Hill. Logs from his and neighboring claims were simply rolled to the strip, where the force of gravity shot them, rumbling and smoking from the friction, down the incline to the mill.

Yesler had managed to build up only a small stock by the time Captain William Talbot sailed over in September 1853, after his exploration of Teekalet to load up with precious millwork for the return trip to San Francisco. But by the following year, Yesler's business was booming to the point where the saws in his mill were wailing into the night to keep up with demand. Local settlers consumed some wood but most of it was shipped to San Francisco — and Yesler even sent one load to Hawaii and another to Australia.

Yesler's place was the center of activity in Seattle — social as well as commercial — and for 13 years the log cookhouse behind the sawmill served variously as Seattle's church, tavern, meeting hall and jail. For the first six years it was also Yesler's home, until he finally brought his wife out from Massillon and built a separate house. The mill hands, numbering 20 in 1860 and making an average of $40 a month, got free board in the big and often crowded cookhouse, where no logger, settler or visitor, with or without money for a meal, was ever turned away.

At one time or another, nearly everybody in town, the local Salish Indians as well, put in a stint at the mill, either regularly or part time. Some of Yesler's helpers became millowners and civic leaders. One of his mill hands, Dexter Horton, left to run the cookhouse for Cyrus Walker at Port Gamble, then returned to Seattle to be a storekeeper. Somewhere along the way he bought a safe—the first in town. People for miles around brought him money and valuables, which he popped into bags tagged with the owners' names. To Horton, who thus got his start as one of Seattle's leading bankers, it did not matter particularly if he remembered the combination. He had picked up the safe cheaply because it had no back. But it looked as good as new standing snug against the wall, and for a long time no one was the wiser.

Unlike Pope and Talbot, who started with Eastern capital and were extraordinarily prudent, Henry Yesler was forever undercapitalized, debt ridden and short of cash. In 1860, at a time Pope & Talbot's Port Gamble mills across the sound were backed by $500,000 in hard currency, Yesler's Elliott Bay operation had a capitalization of only $20,000. And while his mill was the seventh largest of the 25 then at work on the sound, he had trouble keeping up with new milling techniques and buying new machinery to match his competitors'. He himself admitted that his lumber was often cut to irregular lengths and was *cultus*—an Indian word for no good—in other respects. Nevertheless, in 1860, his mill was still the biggest enterprise in Seattle, doing $36,000 a year in sales.

The problem for Yesler, as was often the case on the frontier, was collecting the monies due him and scraping together enough cash to keep operating. Local purchasers, usually even more strapped for cash than he, were slow to pay; it could take months for the money from consignees in San Francisco and elsewhere to make its way north to Seattle.

To meet his payroll one day, Yesler had to borrow $200 from a local minister. Sometimes, when he could not pay his men, Yesler let the workers buy supplies at local stores on his credit—as long as it lasted. He was less trusting of his Indian employees, whom he feared might take advantage of his credit to load up unduly with provisions. Yesler always made sure to pay them in his own ingenious coinage: little metal tabs, stamped with his initials, good for 75 cents or a dollar. That suited the Indians fine: they did not trust Yesler either and wanted something tangible for their labors. Besides, it gave crafty tribesmen a chance to fashion counterfeits on occasion and pass them as the real thing.

For a time in 1860, virtually all of Yesler's mill hands abandoned him to join a gold rush to the Wenatchee River, in the eastern Cascades. Yesler had to close, stilling the saws and the great "gut hammer" gong —an old circular saw outside his cookhouse that was banged at mealtimes to signal the start and stop of work shifts. (Some said Yesler could not afford a proper mill whistle.) But the strike soon fizzled out and the workers straggled back—after Yesler had lost heavily in missed sales. Even worse for him were the periodic gluts afflicting the market. At one point in 1865, Yesler had 500,000 board feet of unsold lumber piled on his wharf, and he complained that the structure was about to collapse into the bay. That may have been true at the time. But over the years, Yesler dumped mountains of sawdust and rock into the bay by his land; eventually his wharves rested on solid fill, and Yesler had added to his real estate a 10-square-block area that Seattle's citizens called The Sawdust.

Despite all his problems, Yesler persevered in 1869 to build a second, much bigger sawmill on his property. For once his timing was impeccable. San Francisco was riding the crest of a silver strike that showered it with prosperity; southern California was just beginning to come alive, providing an increasing market for lumber to satisfy its growth. In 1869 and 1870, Seattle, too, was booming, largely because Yesler himself was pushing to develop its downtown section.

Yesler hung on through all sorts of adversities. When his second mill burned down in 1879, he built a third one, and when that one also was leveled by fire he put up yet another—not in Seattle but on Lake Washington just a few miles away. The next year, in 1889, a terrible fire destroyed 50 blocks of downtown Seattle. This time Yesler was more fortunate. Not only was his Lake Washington sawmill safe, but his new downtown mansion—its roof covered by firefighters with layers of steaming wet blankets—survived the

The Bostons vs. the Klikitats in the battle of Seattle

Seattle's settlers flee to a blockhouse and *Decatur*'s protection in this painting by eyewitness Emily Denny.

In 1855 Seattle was a fledgling town, with 30 houses and 50 white residents, most of whom made a living from the thriving lumber business. Their only concern about the future was the fear of Indians. While the local Dwamish Indians were peaceful, other tribes in Washington Territory bitterly resented the white settlers, or "Bostons," as they called them after their New England origins.

All through the summer of 1855 rumors circulated that hostile Indians, including the Yakimas and Klikitats, were planning to reclaim land by massacring settlers. By October, it seemed wise to take precautions. A volunteer force was formed, a block-house *(above)* was built and the U.S. Navy sent the sloop-of-war *Decatur* to Elliott Bay. To announce its arrival, recalled Emily Denny, who had been a child at the time, *Decatur* "fired off the guns making thunderous reverberations far and wide, a sweet sound to the settlers."

On January 26, 1856, friendly Indians warned them that hostile bands were indeed about to attack. *Decatur*'s captain sent 96 sailors and 18 marines ashore. The party fired a howitzer into the woods, the Indians responded with muskets and the fight for Seattle was underway.

The townspeople, who were either asleep or at breakfast when the battle began, hurried to the fort. One man was so rattled, Emily Denny remembered, that he tried to pull on his wife's petticoat instead of his trousers. Emily's mother was more coolheaded: she took some biscuits from the oven and turned them into her apron on her way out the door.

The fight continued all day, with *Decatur*'s guns booming out reassuring broadsides. By evening, the warriors had suffered a significant number of casualties while the settlers lost just two men; then the Indians retreated. *Decatur* remained on guard through the next summer; but in the face of such firepower, the Indians never again challenged Seattle's beginnings.

Tom Shattuck and Charlie Fick, owners of a sawmill at Boiler Flat, Oregon, whipsaw lumber for nearby gold-mining operations. Such mills remained a part of Western logging up until the turn of the century.

flames. The old man lived to see Seattle grow mightily in his last years. He died in 1892 at 82, a millionaire who had bumbled his way to rare financial status.

The bulk of his fortune was earned not by his sawmills, but by his real-estate holdings in the center of Seattle, expanded gradually from the forested 320 acres on Elliott Bay that he had received for the asking when he first arrived by canoe back in 1852.

Unlike most of the early lumber barons, Asa Mead Simpson, a Maine shipbuilder's son, went West at age 23 as an out-and-out gold seeker. He arrived in San Francisco in April 1850, aboard the bark *Birmingham* after a voyage around the Horn, and headed straight for the sere brown hills and the sparkling ore that had lured him from home. He had good luck in

finding gold—and no luck at all in keeping it. He was back in town within two weeks with 50 ounces of dust worth $800; but a thief made off with half of it, and a bad loan took care of the other half. Simpson still had a 1/32 interest in *Birmingham*—but before long, word came that she had gone down in a fierce storm on her return voyage to Maine.

Yet Simpson was a hardy young man with a lot of Maine gumption. Pulling himself together, he went east of the bay to Stockton, the boomtown at the edge of the mining country. There he opened a small lumberyard and briskly sold off his share of *Birmingham*'s cargo of Eastern lumber. That was the start of a remarkable career that saw him rise to become not only a timber baron but a shipping magnate to boot. And while it was an exaggeration to call Asa Simpson "king of the lumber coast," as some people did at the height of his career, he ran what was without question the most astounding one-man show in the industry.

Like the men of Pope & Talbot, Simpson soon realized that there was no future in importing wood from Maine. In 1851, he began his own survey of the Pacific Coast — with the result that he did for Oregon what Pope and Talbot did for Washington. At Astoria, at the mouth of the Columbia River, he bought an uncompleted sawmill, put it into operation and was soon cutting 15,000 board feet of timber a day, which he transported to California in a small brig. Within a few years, he moved his operations to a much bigger sawmill at Coos Bay between San Francisco and Puget Sound, one of the few harbors on that stretch. By 1858, the mill was producing such a mountain of lumber that it took a fleet of 16 Simpson-owned ships to ferry its output to San Francisco. There was coal as well as timber around Coos Bay; and Simpson correctly figured that when the lumber market slackened, his ships could carry coal to California.

From the start, along with Pope and Talbot, Asa Simpson made San Francisco his company headquarters. It was the West Coast's prime commercial center, the supplier of goods and services, the place to make sales and deals. From there, in the 1860s and 1870s, he managed his sawmill at Coos Bay, and spread out to acquire or build others along the coastal forest belt: at Santa Cruz, Crescent City and Boca in California; at Gardiner in Oregon; and at Knappton, Willapa Harbor and Hoquiam in Washington.

In the early 1880s, Simpson discovered the riches of Grays Harbor, a roomy haven on the Olympic Peninsula of Washington Territory. Somehow, it had been overlooked by the other early lumber lords. Around the harbor, and for miles into the hinterland, grew the massive trees of a great rain forest — thick-boled Douglas firs, Sitka spruces, western hemlocks and redcedars, nourished by up to 150 inches of rainfall a year.

One of Simpson's colleagues, George Emerson, came upon the place on a scouting trip in 1881. He immediately laid claim to an initial 300 acres of timber and hurried back to San Francisco to tell Simpson about the Comstock Lode of the timber world, as it was later called. Next year they moved Simpson's Crescent City mill some 400 miles from California to

the mouth of the Hoquiam River in Grays Harbor and began cutting away at the region's more than 50 billion feet of virgin timber.

Not even Pope and Talbot operated on such a grand scale. Alone among timber barons, Simpson could boast that his logging camps and seven sawmills, strung along 900 miles of coast, supplied his yards with all major Western woods — from pine and redwood to Port Orford cedar, Douglas fir, hemlock and spruce.

Simpson was compulsively expansionist, partly out of ambition and partly because his thrifty Yankee character could not abide dependence on anybody who charged him for anything he could do himself. He refused, for example, to insure any part of his operation — not his mills, his lumber cargo nor the ships — believing that it was cheaper to cover his losses than to pay premiums. "I am now interested in 25 craft, steam and sail, all actively employed with no insurance at all," he once said, and claimed that over the years no fewer than 30 of his vessels had been wrecked, "and I am still largely ahead in my insurance account."

To Simpson, logging and shipbuilding were inextricably linked. He became one of the West Coast's biggest shipbuilders because he hated to buy lumber-carrying vessels that, he was certain, he could design and construct far better than anyone else. There was another reason — a tragic one — for Simpson's intense interest in the seagoing side of the lumber industry. Early in his career, in 1856, he had sent for his brother Louis to come West, intending to have him run the big new sawmill he was putting up at Coos Bay. But the vessel on which Louis was arriving struck a sand bar at the entrance to Coos Bay and broke up in heavy waves, carrying him to his death. Similar treacherously shifting sand bars guarded the mouth of virtually every river and harbor in the Northwest, which was why Pope and Talbot had opted for Puget Sound, placid and deep and free from such hazards. What was vitally needed south of the sound, as the loss of his brother hammered home to Simpson, were strong, steam-powered tugboats and knowledgeable pilots to shepherd vessels safely in and out of the lumber ports.

In 1858 Simpson bought the sturdy teakwood brig *Fearless,* reconstructed her as a steam tug and sent her up to Coos Bay, where she helped pioneer the West Coast tugboat business, towing his own and other lum-

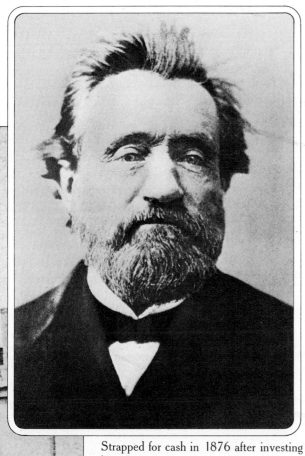

Strapped for cash in 1876 after investing heavily in real estate and other ventures, lumberman Henry Yesler concocted an audacious scheme to get himself out of debt. He promoted a lottery of 60,000 tickets, sold at $5 apiece, offering 5,575 prizes — mostly small; but the grand prize was to be Yesler's sawmill, at far left in this view of Mill Street in Seattle. There were two flaws in his thinking: the area that Yesler canvassed contained barely 6,000 inhabitants, and the sawmill was mortgaged to the hilt. He was saved from immediate ruin when a local judge ruled the lottery illegal and ordered the ticket money returned.

ber ships clear of danger. For a time, Simpson skippered *Fearless* himself, along with other tugs his new shipyard at Coos Bay turned out. One of his biggest jobs was towing ships past the great bar at the mouth of the Columbia River. Over the years the bar had claimed so many victims that it won the nickname the Graveyard; the state of Oregon promised a $30,000 subsidy to anyone providing tugboat service for five years. Simpson took up the offer and assured safe passage with his powerful tug *Astoria,* collecting large towing fees as well as the subsidy.

Once Asa Simpson had fallen in love with shipbuilding as an adjunct to his logging operations, his versatility and inventiveness seemed endless. He built dozens of lumber carriers in his Coos Bay yard alone. Many were little flat-bottomed steam-and-sail schooners — vessels that could nimbly bob their way in and out of the tiny, wave-lashed California "dog holes" *(pages 72-73),* the coastal indentations so small, people said a dog could hardly turn around in them.

If Simpson's lumber vessels were agile, they were also extremely strong. Curved-keeled craft, he had noted, tended to break their backs when they grounded on sand bars while carrying heavy loads of boards. He wanted his boats to "have a long straight part on her bottom to rest upon in case of grounding with a full load on." From his shipyards came America's first four-masted barkentine, and its first five-masted schooner, both launched as lumber carriers. The schooner design was fore-and-aft, with triangular instead of square sails. A schooner could not catch as much wind as a square-rigged ship, and thus was at a disadvantage for long-distance voyages; but for shorter runs — coastwise or to Hawaii — it was more maneuverable — and manageable by a smaller, more economical crew. Simpson's schooners were gems of their kind and in the 1870s and 1880s, at the height of his fame, he sold nearly as many ships to others as he built for his own fleet.

In every port between Puget Sound and San Francisco, Asa Simpson became a familiar figure and, as the years passed, he was known as much for his eccentricities as for his power and wealth. Nobody ever saw him in shirtsleeves. He always dressed formally — frock coat, wing-tip collar — and he affected a silk hat, which inevitably brought him the nickname of Stovepipe Simpson. (Once he fell overboard from one of his lumber ships during a storm; before he could be hauled back aboard, he splashed about until he had rescued his beloved hat.) He was also crotchety and got into epic arguments with associates who attempted to change mill operations or alter ship designs on their own. He once removed a mill manager who had the temerity to replace inefficient Chinese laborers with more highly paid but more productive Americans without first consulting the boss. And the son of one of his master shipbuilders recalled a bitter dust-up over, of all things, the placement of the ship's head on a vessel under construction. Simpson wanted the toilet located under the forecastle deck instead of in the forward deckhouse. The builder protested that the deck was too low and Simpson, naturally, undertook to prove him wrong. Recalled the son: "Squatting down, he backed into the space but was unable to lower his trousers. He crept into the open and stood erect to half-mast his nether garment and again backed under the forecastle. This time his head hit a deck beam with a resounding crash." Not in the least daunted, Simpson still insisted on putting the toilet under the forecastle — except that now he directed the builder to cut a notch in the deck beam so no more heads would be cracked on it.

Though he was certainly a multimillionaire, there is no record of the exact extent of his wealth, or even the grand total of his properties and investments. The fire that followed the San Francisco earthquake of 1906 and another that leveled his son's Coos Bay mansion a few years later destroyed the bulk of Asa Simpson's records. His empire was at its zenith in the late 1880s and early 1890s. After that, the coming of railroads relegated his fleet of coastal lumber vessels to a secondary role. As his sawmills grew older and less efficient, he was reluctant to modernize or move them to better locations near railheads. And finally, giant corporations arrived to manage the woods and doom the old-fashioned one-man empire. In the end, little of his fortune remained. But Simpson himself outlived all

the first generation lumber barons, to die in 1915.

John Sutter, Pope, Talbot and Walker, Yesler and Simpson were major players in the great drama that quickly unfolded with the frontier's final thrust west to the Pacific Coast. But there were hundreds of other lumbermen who influenced history in one way or another.

Some played historic roles, like Baltimore skipper Stephen Smith, who in 1844 unloaded machinery from his *George and Henry* near Bodega Bay, California, and put up the very first steam sawmill on the West Coast. Some played incidental roles, like New Englanders Francis Pettygrove and Asa Lovejoy, who stood on a bank of the Willamette River one day in 1845 and tossed a coin to see which of their hometowns would be honored in the name of the new city being platted there; Pettygrove won and it became Portland, instead of Boston, Oregon. Some men played minor dramatic parts, like Captain James Ryan, who rammed his steamer *Santa Clara* ashore at Humboldt Bay in 1852 to use its power for a sawmill he wanted to build. The ship's engines drove the saw and the grounded vessel served as bunkhouse and cookhouse. Ryan watched three ships in a row carrying lumber from his mill break their backs on the Humboldt bar before a fourth made it to market in San Francisco.

And who might be the necessary villain of the piece? That rich — and rewarding — role was filled by Henry Meiggs, unforgettably cast as the timber baron turned scalawag. A portly, 38-year-old lumber dealer from New York City, Meiggs hit San Francisco in the summer of 1849, bringing around the Horn a shipment of boards that he speedily sold for 20 times what it had cost him, making a clear profit of $50,000. No man to ruin his manicure by scrabbling for gold in gravelly stream beds, he bought land at North Beach, where he thought the city's future lay, and put up a steam-powered sawmill at the water's edge. To feed the mill he rounded up a small army of about 500 drifters and sent them across San Francisco Bay to cut down a

Harry Meiggs, one of San Francisco's first big lumber entrepreneurs, skipped town in 1854 leaving a trail of bad debts. He later surfaced in South America and became a millionaire banker and railroad builder.

magnificent grove of redwood trees, which were then floated back to his mill in large rafts. Reputedly he made a quick half-million dollars from that single venture. Soon he expanded north into the redwoods and pines of Mendocino County, where his California Lumber Company built one of the largest and most productive sawmills in the state, with a capacity of 50,000 board feet a day. But with orders for twice that amount, a second mill had to be quickly added. The money came rolling in.

Handsome and invariably hailfellow, "Honest Harry," as Meiggs became known in San Francisco, was always good for a touch from a friend or stranger down on his luck. A dedicated music lover, he helped found the city's philharmonic society and erected a music hall on Bush Street. As a civic-minded alderman he laid great plans for the city—especially the remote North Beach area, where with the profits from his lumber operations he bought large tracts and built an enormous pier. Eventually, land speculation turned out to be his undoing. In 1854 San Francisco suffered an economic depression and Harry Meiggs, with no buyers for his lots and a poor market for his lumber, found himself up to his ears in debt. He borrowed heavily, but taxes and assessments, property-development costs and a whopping $30,000 a month interest on his loans soon forced him into bankruptcy.

In October 1854, he skipped town on the bark *America,* taking along his family, two servant girls and an ample stock of fine wines. The ship was well underway for Tahiti before his fellow citizens realized that their genial alderman had fled—and learned the reason why. The scandalized *Alta California* reported: "San Franciscans are reluctantly convinced that Henry Meiggs is not only a bankrupt, but a forger, cheat and gigantic swindler." That was an understatement. Over a period of months Meiggs, who as a trusted alderman had easy access to City Hall and its casually kept books, had stolen a large quantity of blank city warrants — issued like checks in payment for city services — and

fraudulently cashed or pledged them as security all over town to the tune of some $800,000. His disappearing act triggered a financial shock wave that rocked the city from one end to the other.

Meiggs tarried only briefly in Tahiti before surfacing in South America. His huge expenses in San Francisco had evidently eaten up all except about $8,000 out of the $800,000, and he was soon so broke, it is said, that he was reduced to pawning his gold watch. But having hit bottom, Meiggs suddenly shot to the top again, this time as a big-time railroad builder. Though he had no engineering knowledge, he was an experienced contractor who knew how to deal with customers, suppliers and workers in the field. Employing the very real Meiggs nerve and charm, he made a new fortune building railroads in Chile; and in the 1870s he won multimillion-dollar contracts from the government of Peru to construct two railroads over the Andes. One of them went from Arequipa to the town of Puno on Lake Titicaca and crossed the mountains at 14,665 feet; the other, part of a grand scheme to connect the Atlantic and Pacific oceans, traversed a pass at 17,000 feet. Meiggs fulfilled contracts to build hundreds of difficult miles of other South American rail lines. He made millions—as much as $100 million by some reports. Eventually, he paid back most of the money he had embezzled in San Francisco, and in 1873 the California legislature voted to exonerate him, in effect saying, "Come back, Harry, all is forgiven." But the governor vetoed the amnesty, and when Meiggs died four years later, he was buried at Villegas, his estate near Lima, where he was mourned as the Railroad King of the Andes.

No other timber baron had ever fallen so low or rebounded so high. And even in his exile Honest Harry was a factor in the California lumber industry. The tracks that carried his railroads over the high Andes rested on ties of tough California redwood—millions upon millions of feet of it—ordered by Meiggs at top dollar and shipped from his old mills at Mendocino.

The art of loading the dog-hole schooners

The sight of seven schooners crowding the rock-strewn shoal waters of a tiny inlet on the Pacific illustrates a paradox of the early West Coast lumber industry. Between San Francisco and Oregon, the coastline bordering one of earth's richest forests had few harbors worthy of the name. And yet there was no way to transport lumber south to market except to carry it, by Herculean effort, over rocks and heaving waters onto the custom-built lumber schooners. There were no railroads from the mills to the markets; and the sea, how

ever difficult of access, was the only highroad open to lumber barons-to-be.

To make use of it they built compact, two-masted schooners capable of entering the coast's harbors—called dog holes, presumably because some salty early skipper found them barely big enough for a dog to turn around in. Moving the timbers from their mills to these agile vessels, lumbermen performed prodigies of improvisation and labor, as shown on the following pages.

The lumber was hauled overland by the shortest possible route to a point

on the shore. There they built wharves on trestles across the bluffs and rocks. From the ends of the wharves they slung inclined wooden chutes, down which the boards slid, often with perilous momentum, onto moored vessels. In the 1870s, when high-strength wire cable became available in California—to haul streetcars up and down San Francisco's hills, among other things—a number of more sophisticated and flexible loading devices were invented, adding another colorful phrase to the logger's lexicon: loading "under the wire."

72

Whitesboro, a lumber schooner with steam power, loads from a bluff-emplaced wire "chute" at Greenwood, Mendocino County. A cart of lumber was let down and hauled up the incline by cable from a steam-powered winch. A mule pulled the cart along the track to the loading trestle.

75

Nestled in a wire sling beneath a pulley-wheeled "traveler," a load of boards rides down a cable to a lumber schooner at Bourn's Landing in the southern Mendocino County port of Gualala. This particular wire system was copied from the rigs used by gold miners to remove ore from Sierra mines.

Two anchored lumber schooners ride out a storm off Mendocino City, while a boatload of seamen struggles for headway in the huge seas. Coastal storms were frequent, and the toll of ships and men over the years was in the hundreds. On November 10, 1865, alone, 10 schooners were lost.

3 | Taming the virgin forest

The awesome size of the Western conifers called for logging techniques on an equally heroic scale. A single Douglas fir could provide as many as four 32-foot logs, sometimes more than 10 feet thick and weighing up to 100 tons. Simply to move such a mass of timber from stump to sawmill was a challenge to the lumberjacks' energy and ingenuity. They devised all sorts of chutes and flumes to carry timber down from the hills, and built elaborate skidroads along which teams of horses and oxen dragged strings of logs.

In the 1880s, new machines, notably the steam-powered donkey engine *(pages 106-109)*, changed logging for all time. In a high-lead setup *(below)*, with its steel cable running through a pulley near the top of a towering spar tree, a huffing donkey could yank the mightiest log out of the woods in minutes, and a sharp crew could handle 5,000 tons of logs in eight hours.

A high-lead rig in Washington brings in a pair of logs and some well-balanced joyriders.

A freshly felled log, evidencing the smoothness of a sawed back cut, lies at rest after being hauled out of the woods by the steel cable of a donkey—one of nearly 300 used by Washington loggers around the turn of the century. The large hooks were used to align the log in a storage area.

82

An ox-drawn wagon hauls pine logs from an Oregon forest about 1880. The wheels are thick slices of wood rimmed with heavy metal bands. Such wagons were especially useful in areas where crude logging roads—little more than parallel ruts—could be cut through relatively level timberland.

85

After descending along a wooden chute in California's Sierra, logs are eased down an inclined rollway and loaded onto the flatcars of a horse-drawn tramway for transportation to the mill. The tramway's rails were made of wood and reinforced with strap iron to support the heavy loads.

Grand schemes and Herculean machines

Clem Bradbury was new to the logging camps of the Western forests. He was a Maine man, from York County, a sturdy, self-reliant lumberjack who at 28 had spent most of his working life hewing white pine in the woods of his native state. But now, in January 1847, he was in Oregon country, within sight of the lower Columbia River, and he was about to lay his ax against a tree the likes of which he had never encountered before. The trunk soared straight up, nearly 300 feet into the sky. At the base, the bole swelled outward until it was eight feet thick. The tree, an immense Douglas fir, had been standing there for centuries, perhaps 1,000 years.

Bradbury fell to. But sinking his cut barely 16 inches, he halted in astonishment. From the wound rushed a fountain of pitch—barrel upon barrel of crude turpentine. To continue was impossible; Bradbury put down his ax and that night in the bunkhouse he said very little. By morning, however, the heavy flow of pitch had ceased and Bradbury resumed chopping at the same tree. At last, after six hours of back-breaking labor, the fir groaned and thundered to the ground. Back home, Bradbury could have felled half a dozen white pines in the time it had taken him to conquer this one Western giant. And similar specimens grew all around in a thick, incredible, evergreen carpet.

Clem Bradbury had learned the first and most important lesson about Western logging: things were very different out beyond the Missouri. Nothing in the vast pine stands of New England or the Lake States could prepare lumberjacks for the awesome trees they confronted in the Pacific and Rocky Mountain forests. In the early days the difficulties and challenges these strange woodlands presented stretched to the outer limits the loggers' capabilities—all their great muscles, agility, ingenuity and courage. So, over the years, they devised an array of amazing new techniques and fantastic machines to bring down the monsters and, having felled them, to carry them off to market.

Though they were fiercely individualistic and would remain so for all time, loggers in the West soon realized that teamwork was twice as important as it had been back East. At first a logging crew consisted of perhaps a dozen men, who took their orders from the "bull of the woods," or boss logger. It was he who, at six in the morning, bellowed "All out for the woods!" to signal that breakfast was over. And it was he who held the ultimate responsibility for deciding which trees to fell, how they should fall, and where the cuts for logs should be made once they had fallen. After that, it was the fallers and buckers who transformed the towering forests into the logs that were moved to the mills.

In the East, one man equaled one tree. Out West, the choppers or fallers (never called fellers, even though their job was to fell trees) almost always worked in pairs and seldom attacked a big tree at ground level. It took two men to cope with the great girths of the larger trunks; and they cut high for several reasons, as newcomers like Clem Bradbury soon discovered. A tree standing on the side of a hill, for example, offered unsafe footing for a faller, so a platform of sorts had to be employed. A tree growing out of thick underbrush was hard to reach with fast-swinging axes; if the underbrush could not easily be cleared, it had to be avoided. A swollen base was too time-consuming to chop through; moreover, pitch collected there, and the bole was often riddled with less-than-perfect wood, the fibers having

Workers lay a cradle of cross ties for a skidroad in California about 1880. Once the ties were in place, the cradle was partially filled with earth, and logs could then be yanked down the road by cables from the steam donkey engine in the background.

been stretched and loosened over the years by the swaying of the trunk.

So the two axmen—head faller and second faller —plied their trade perched high off the ground. Some teams built elaborate scaffolds to stand on, 12 to 20 feet up the trunk. However, most loggers preferred five-foot-long by eight-inch-wide springboards (page 118), platforms inserted right in the trunks of the trees; they were a lot faster and easier to put up and worked almost as well. Each faller faced his partner across the front of the tree, cut a narrow notch in the side of the trunk nearest him, inserted his board and climbed onto it. If he wanted extra height, he would hack a new perch as high up the tree as he could reach, and then move to the higher position. There were two ways of doing this. The usual method was for each man to use two or three springboards and position them like steps up the trunk to the proper height. When the tree began to topple and the logger made his hasty descent, he could hop more or less deliberately from board to board —a total drop of perhaps 20 feet.

An especially athletic and flamboyant faller might choose a more dramatic method, which employed only one springboard per man. Having made the higher notch from his first off-the-ground position, he would drive his ax deep into the trunk above him and cling to its handle with one hand while repositioning his springboard with the other. When it was firmly in place, he would haul himself up by the ax handle and swing aboard with a flourish—repeating the action if he needed still more height. Dropping down from the tree on only one springboard meant that the logger had to take off in a flying, thudding leap to earth.

It took two cuts—an undercut and a back cut—to send a Western monarch crashing to the forest floor. Facing each other on their perches, one man swinging his ax right-handed and the other left-handed, the fallers fashioned their single undercut with sculptor's care —flat on the bottom, angled at about 45 degrees on the top. The incision reached more than halfway through the tree and was carved as smooth as if it had been planed. Next, the fallers moved their springboards to the opposite side of the tree, and made a flat, horizontal cut only deep enough to rob the tree of support, so that it leaned and then fell toward the undercut side. In the woods a faller was known for his accuracy no

less than for his strength and stamina. Lives—his as well as those of his mates—rode on that skill. When the fallers hollered "Timber!", tossed their axes into the brush and hurtled down from their springboards, they had to be sure the tree would fall true, in a direction determined by the placement of the back cut in relation to the undercut. If an expert faller was feeling good and looking to impress someone or win a bet, he would, before commencing his cuts, plant a stake in the forest floor 200 feet away from the tree. If all went exactly right, the upper end of the trunk would descend on the stake and drive it straight into the ground.

The entire process, involving notches and springboards, undercuts and back cuts, was a Western improvisation. Taking down a big fir or cedar or a fat sugar pine could occupy two men for a whole day. Even a Douglas fir five feet in diameter took about two hours of undercutting plus half an hour of backcutting. In California's redwood country, chopping through a 20- to 25-foot trunk could mean a week of hard labor for a pair of woodsmen. An alternative method was to undermine the tree not by chopping but riddling it with auger holes, a procedure that in the early days could take several weeks. In 1853, five men new to the giant sequoias of the Sierra spent 22 days boring holes with a long auger all around the 26-foot-thick trunk of a 300-foot-tall specimen in the Calaveras Grove. Large steel wedges were then inserted into the auger holes and driven deep into the tree by battering-ram logs suspended from the trunk. One noontime, when the crew was at lunch, a stiff gust of wind finally toppled the monster; since few sequoias had ever been felled before, people for miles around the area assumed an earthquake had struck.

In later years, redwood loggers tried stuffing sticks of dynamite into the auger holes. The charges would then be set off simultaneously, blasting the giant off its base. The method was spectacular but sometimes wasteful. In the absence of carefully angled cuts, the direction of fall was impossible to control. Furthermore, a redwood, being brittle, could shatter into useless fragments on impact with the ground. Lumbermen preferred the long, sweaty but more precise way—chopping—to direct the fall and protect the tree. A crew would spend days preparing a long soft bed of boughs, known as a layout, to cushion the falling redwood. Once the tree was down

intact, loggers did use dynamite to prepare it for the sawmill. Until well into the 1880s, the saws available to the mills were too small to slice through the vast diameter of a Sierra redwood trunk. The customary, laborious way of handling the biggest trees was to saw the great carcass into logs and then blast each log in half lengthwise with dynamite charges. Sometimes the explosion fragmented and wasted the entire log; with luck, however, it split cleanly into two half sections, which could then be coped with by the mill saws.

In the late 1870s, West Coast fallers discarded their traditional, single-bit, Eastern-style axes for a new type introduced by the Washington Mill Company of Seabeck: concave-headed, double-bit axes, weighing three and a half pounds, with handles made in Pennsylvania from strong, supple, second-growth hickory. The double bit was an advantage to a woodsman who needed a sharp cutting edge, but who sometimes had to cut material—such as dry knots—that quickly dulled the blade. In effect, one blade spared the other, and thereby spared the faller from carrying two axes. The handles were straight, 38 to 42 inches long, so that either bit might be used with equal ease.

"Strangers," observed a Humboldt County logger, "even those familiar with lumber chopping in the East, might wonder why such long handles are used. It is all on account of the size of our trees. The woodman must be able to reach the center of the tree from his position on the platform, and it must be remembered that in many of our redwood trees the center is four to six feet from the outer edge." That was an understatement. No ax was long enough to reach the center of many Douglas firs, let alone the two species of redwoods, and fallers actually had to stand inside the undercut to extend it to the heart of the trunk. This was backbreaking work and, as the same laconic logger commented, "a man that swings over three pounds of steel all day at the end of a 42-inch handle must be possessed of much power and endurance."

Considering the brutal labor involved, it is surprising that it was not until the 1880s that fallers set aside their axes and began using two-man crosscut saws to make their second back cuts. For nearly 40 years, right before the fallers' eyes, the "buckers," or log sawyers, had wielded labor-saving crosscut saws to slice felled trees into logs. Finally, it dawned on the lum-

bermen that although the undercut, being a notch, had to be chopped out by ax, the straight back cut could be sawed. When the fallers were finally equipped with two-handled, longer versions of the crosscut saws, they found that their work went much faster and more safely; a tree that had been sawed instead of chopped opposite the undercut was less likely to twist as it toppled.

There was something noble in the sight of two fallers, swinging or sawing rhythmically up on their springboards. But the lonely bucker—down deep in the brush, pushing and pulling his saw through an enormous trunk—was a portrait in sweat-drenched misery. His job was to cut the felled tree into logs, usually 16, 24, 32 or 40 feet long—the lengths preferred by the mills. Approaching a downed trunk, the bucker would measure it off with an eight-foot stick. Then he cleared away enough room in the surrounding underbrush and went to work. Except on the enormous redwoods, the bucker worked alone, for his pace was his own and two men seldom worked harmoniously on a single saw. ◉

Tools and trappings of the timberman's trade

The woodsmen who arrived in the West from the Eastern and Midwestern forests brought their axes, saws and peavies with them. But challenging terrains and trees of unprecedented proportions taxed some of the old tools. New techniques had to be invented, and with them new implements.

East of the Cascades or the Sierra, few trees required a falling saw more than six or seven feet long (the girth of the trunk plus two or three feet of travel for saw strokes). But western red-cedars and Douglas firs often required 10-foot saws, and the redwoods and giant sequoias took blades up to 18 feet long. Loggers never before had to make undercuts eight to 10 feet above the ground to get above a tree's base flare. The invention of the springboard solved that problem; the springboard, in turn, required a new kind of ax. To cut a notch in a trunk that was deep and narrow enough to hold the springboard's pointed tip firmly, an axhead was needed narrower and longer than the old 4½-by-9-inch blade. And so the Western falling ax was born, with a double-bitted head more than a foot long and 3½ inches wide.

The sawyer's second most important tool was the container for his saw lubricant, which thinned the pitch that spilled on the blades; it was always called oil, although it usually was kerosene. Oilcans were available, but most loggers preferred a quart whiskey bottle; they could then claim that they needed a new one for each quart of oil.

LUBRICANTS
Lubricants were carried in cans or bottles fitted with hooks so they could be hung on branches. Thin spouts or notched corks emitted the lubricant in drops.

FALLING SAW
The faller's crosscut was long (like this 10 footer) and narrow, so it would sink quickly into the tree and let anti-binding wedges be driven into the cut.

FALLING AX
The Western ax, with a longer, narrower blade than Eastern axes, had a longer haft — up to four feet — to reach into the heart of trees eight feet in diameter.

BROADAX
Most useful in the field, where there were no mills, the broadax had a chisel-like blade to square logs in trestles and flumes.

SPRINGBOARD
The tip of this faller's perch had a metal V-lip that dug into the upper wood of a notch in the tree trunk and thus wedged the springboard firmly in place.

BUCKING SAW
The bucker's crosscut was used for cutting up felled trees. This model was designed for use by either one or two men.

PICKAROON
This tool's sharply pointed head
was driven into a log to "horse" it—i.e.,
to move it by sheer muscle power.

WOOD MALLET
With a 36-inch haft and a hardwood head
five inches in diameter, this sledge was
used to drive wedges into sawcuts and also
to help split shingles and shakes.

WEDGES
Driven into sawcuts to spread them and
keep blades free, these tongues,
made of steel or wood, were also used to
control the direction of a tree's fall.

BÖKER
This wood-and-metal jack, named for its
German maker, was employed in the early
days to move logs too large to be
manhandled. Böker levers, reversing when
their locks slipped, broke many an arm.

CALKED BOOTS
Pointed metal calks, which loggers pronounced "corks," gave purchase on logs, springboards and rough terrain. They came in various lengths for different jobs.

CANT HOOK
A five-foot pole with a hinged, toothed hook on the end, this tool was used to roll logs on the ground or on mill landings.

PEAVEY HEAD
A Maine-born tool used to maneuver logs on land or water, the peavey had at its tip a log-prying spike that was lacking in the cant hook, which it came to replace.

CLIMBING IRONS
This sharp-spurred climbing aid fit under the foot and was strapped to the leg at the ankle and below the knee.

UNDERCUTTER
When a log's position made it necessary to saw it from underneath, this tool's chisel-like end (*right*) was driven into the underside up to the curved detent; the saw, teeth up, rode the grooved wheel.

95

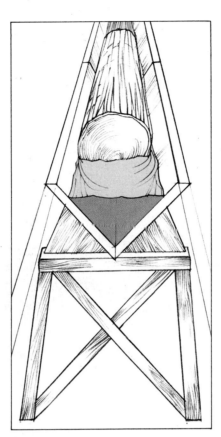

A V-shaped flume provided a rapid means of conveying wood, according to inventor J. W. Haines in 1870. It was superior to flat-bottomed flumes because it required less water and was less prone to log jams.

To keep his saw from binding, the bucker carried with him wedges of steel or wood to hammer into the cut as it deepened. He brought along a water jug to slake his thirst and a bottle of saw oil with a hole in the cork; when the saw ran into pitch and stuck, he simply sprinkled the blade with the coal oil or kerosene from the bottle. Bucking was hazardous work. The bucker had to be constantly concerned not only about the dangers overhead, but also about the prostrate, seemingly innocuous trunk he was attacking on the ground; when severed, it might lurch as the tension of its many tons was suddenly released by his saw.

Fresh from Sweden at the age of 16, a logger named Emil Engstrom hired out as a lonely windfall bucker on the lower Columbia River in 1903. A windfall was a tree blown down in a storm, and it had to be cut apart in order to get it out of the way of trees being purposely felled. "I was given a seven-foot-long bucking saw," Engstrom recalled, "a quart bottle full of saw oil, an ax, a 10-pound sledge hammer and four seven-pound steel wedges—a full load for a strong man." For some unknown reason, said Engstrom, "windfall buckers were always given the worst saws there were in the buckers' saw rack; some were kinked and others had a few teeth missing. The bull of the woods took me out and put me to work on a nice pile of windfalls, some of them seven feet at the butt, far out in the standing timber to work there alone. I managed not to split a log the first few days.

"In some of the big butts I struck pitch. Saw oil has little effect on pitch when it is running out each end of the saw cut. A slow stream of water, if water is handy, will do some good. Then, too, timberbind in a tree is an evil hard to understand. Even if the bucker can drive a wedge or two, the cut may snap his saw, so there is no way out of it except to chop the saw out."

Even more of a challenge than getting the monsters down and bucked to size was the prodigious task of hauling them from where they lay. In the woods of New England and the Lake States, logs were moved to the mills by horse-drawn sleighs or floated in rafts across lakes or down networks of rivers. But out West the land was mountainous, the undergrowth dense and usable rivers were nowhere numerous.

Trees that grew at the water's edge—on the Columbia River, Puget Sound or Grays Harbor—were not a problem; they could simply be felled and floated away. But the forests extended many leagues inland from the nearest shore, into vast stretches of wild mountain country. An incredibly rich resource grew there for the taking. But the timber was no good just lying on the forest floors; it had to be moved, often over great distances, to the mills and markets whose appetites were becoming insatiable. Getting it out—by somehow coping with the gargantuan hulks and the rugged terrain—meant developing a whole new approach to logging. Sheer brawn was no longer enough. But it helped; and in the beginning, before a series of explosive technological advances invaded the woods to transform Western logging from a crude handicraft operation into a great industry, the loggers coped by using a combination of brute strength and their brains.

On a midsummer morning in 1854, the woods along Puget Sound were all but silent. Droplets of fog, condensing on evergreen branches, fell soundlessly onto the soft carpet of fir needles on the forest floor. From somewhere deep in the trees came the faint hammering of a pileated woodpecker's sharp bill, chiseling fat slivers of wood out of a massive trunk. A mile and a half away the saws of the new Pope & Talbot sawmill at Teekalet were biting through logs, but their whine was audible only when breaths of an offshore wind riffled the treetops 200 feet overhead. Up there, curls of mist

96

Cut through rock and lined with wood, this spectacular dry chute was used to send logs from the hills to the Klamath River in Oregon.

thinned and evaporated in patches of blue sky; it was a day when the sun would burn off the fog.

A sudden clanking of metal broke the silence. A loggers' skidroad passed this way — a wide but unobtrusive trail intermittently paved with timbers laid crosswise; and around a bend in the road appeared a double line of oxen, enchained like slaves. A lean and bearded bull whacker — the slave driver — in floppy hat, galluses and calked boots, walked beside them bawling curses. "Hump, you, Buck! *Move!*" he roared, and stabbed at their rumps with his goad — a stick with a sharp nail driven through the end. He had eight yokes, or pairs, of reluctant animals to manage, all straining and grunting, their hooves pawing for the road's timbers and plopping heavily into the mud between each one.

Behind the beasts, sliding and yawing, came a chained-together string of a dozen enormous logs, each one five feet thick and weighing close to 10,000 pounds. The sounds of clanking, cursing, grunting and jolting gradually increased to a din as the train moved near. Then the racket faded as the oxen, flicking their tails to keep flies at bay, disappeared around another curve with their "turn" of logs in tow. Fifty more tons of timber were bound for the saws at Teekalet; and as the cargo went its ungainly way the forest's accustomed quiet returned.

The skidroad made its appearance on Puget Sound in the early 1850s. The name of the man who invented it is lost to history, but his idea was a brilliant one for its time. Instead of just nibbling away at the fringes of the forest, lumbermen could now take trees growing a mile or so farther into the woods. But the skidroad had to be meticulously engineered. The roadway had to lead generally downhill, toward the shore — though, for the sake of the oxen, not too precipitately. Its curves had to be gentle and banked, to help the logs along. "Swampers" cleared the route with axes, mattocks and shovels. Buckers prepared the skids: foot-thick timbers not less than 12 feet long, cut from whatever trees were handy. The skids were laid crosswise, like railroad ties, and half-buried in the ground at seven-and-a-half-foot intervals so that a traveling 16-foot log, the shortest standard length, would always rest on two of them. A scallop was cut out of the top surface of each skid to cradle the passing logs; when the roads were made out of softwood and intended for long use, a four-inch lining of hardwood, usually maple, oak or madrona, was precisely fitted into the scallop to reinforce the skid.

Every log had a "ride" of its own, a natural position in which it rested whether floating in water or lying on the ground; a good hooktender — the man charged with getting a turn of logs ready for skidding — made sure that each one lay ride-side down. The forward end of each log was "sniped" — rounded and smoothed with an ax so that it would not tear up the roadway as it hit the skids. The couplings between logs were metal chains, held fast by bent metal spikes called dogs. A long chain ran forward from the lead log, passing between the pairs of oxen, and was attached to the heavy wood-and-leather yokes that hung over the animals' necks. The kind of yoke was important: the best wood for yokes was said to be Bishop pine, because the pitch it contained seemed to toughen the animals' necks and help prevent gall sores.

The man who made it all go — the engineer of the log trains that dragged down the skidroads — was the bull whacker, or "bull puncher." Together with the bull of the woods, he was the top man in the forest, making $100 a month when ordinary loggers made only a third of that. He was in absolute charge of the teams of 1,800-pound Ayrshire or Durham oxen. These were valuable beasts; a pair of leaders — the front yoke — could cost up to $300. And it took as much as a month to teach a new animal to move ahead, stop, pull hard or easy and turn left or right in response to its name, the bull whacker's cussing or the sting of his goad. He fed them hay and bran, groomed and occasionally bathed them, rubbed arnica into their chafed necks, nursed them when they came up lame, fitted brass caps on the sharp ends of their horns and, on Sundays, helped a blacksmith shoe them.

Assisting the bull whacker as he guided his charges down the skidroad was a lad variously known as the skid greaser, swabber, grease dauber or water boy. He was generally a teenager learning to be a logger, and it was his thankless duty to scamper just behind the rearmost yoke of oxen and just ahead of the front log, swabbing the skids. He used whatever lubricant was available — fish oil, mutton tallow, bear grease, tubs of old butter or water — to reduce friction and help keep the logs moving. Where the downgrade was steep, he

With a huge log slung from the axle, a pair of "big wheels" is ready to roll away to an Oregon mill in the 1880s. An Eastern device, big wheels worked in the West only in relatively flat, dry, inland forests.

would throw dirt on the skids to slow down the logs and keep them from piling into the ox team. The bull whacker did his part by looping heavy chains around the logs to dig into the road and act as brakes.

At the end of each trip of an hour or two, the skid greaser's job was to gather up the chains and dogs, and pile them into a "boat" for the return to the head of the road. The boat, which rode at the tail end of the log train, was a log hollowed out to hold jack screws, levers, axes and extra chain. The climb back uphill, with only the boat in tow, was easy for the oxen and for the lowly skid greaser. He had to sweep debris off the road, but for brief spells he could ride in the boat while catching his breath.

For all its ingeniousness, the skidroad had its limitations: it could not work on a hill whose gradient was too steep for the oxen, and it could not be more than a couple of miles long because not even the doughtiest of animals were strong enough to pull the heavy logs very much farther. Yet some of the best timber in the West grew in mountainous terrain too steep or too far into the back woods for skidroads and oxen to be of much use in transporting it. Loggers wracked their brains for fast, effective ways to move the big sticks down from the heights. The log chute, a spectacular device, was the first that did not rely primarily on the muscle power of men or beasts.

The principle behind the chutes was simplicity itself: since the course was downhill, let gravity do most or all of the work. In essence, the chutes were long troughs, usually made of peeled tree trunks, that served

99

Planks of milled lumber float downhill end to end in an Oregon flume. Such water-filled troughs generally moved roughly cut wood from small sawmills up in the hills to larger centrally located company mills, where it was planed smooth after its battering descent and cut into salable sizes.

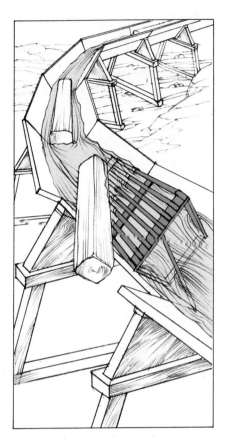

as conduits for the logs. No one thought to record when they first were put into use, but they are known to have been in existence in the Far West by the late 1850s. Some chutes inclined only gradually and were kept well greased to facilitate the passage of their burden. This version had a towpath alongside it on which teams of horses traveled while easing the logs on their way. That sort of chute was simply a smoother-surfaced, somewhat steeper version of the skid-road. Another kind of log slide, faster and much more dramatic, was the steep "running" chute, which worked entirely by gravity, carrying logs down precipitous mountainsides at dizzying speed, to pile up at a landing or plunge into a pond at the bottom.

One of the most famous running chutes was built high above the Klamath River, just north of the Oregon-California border, by a logger from Michigan named John Cook. His men were cutting sugar pine in the Cascade mountains in 1891 and his problem was to get the logs down to his mill on the river at Klamathon, nearly 3,000 feet away. Cook's solution was a steep trough of smoothly hewn, greased timbers. It dipped through a cut blasted in one intervening hill, crossed a depression in the land on a trestle, passed through another cut in a second hill, and ended at a roomy pond that opened on the river. Cook boasted that logs sent down his 2,650-foot Pokegama Sugar Pine Lumber Company chute made the trip in 22.5 seconds, which came out to 90 miles an hour. They arrived smoking from the friction and hit the river with a tremendous, sizzling splash. They were instantly cooled off, but they often smacked into the pond so hard that the head ends split wide open.

John Cook's enterprise was active for a profitable decade, until a fire in 1902 wiped out not only his company and his sawmill but the little mill town of Klamathon as well. Cook's running chute remained, a thin

scar drawn on the mountainside and a monument of sorts to its inventor.

Chutes were a wonderful rapid-transit device as far as they went, which could only be down one or two slopes. But there were many places where swift, long-range lumber movement was needed and the loggers' answer was the flume — an equally sensational enlistment of gravity and surely the most outlandish overland-transport device ever seen, West or East. Built of boards and fed with water from a reservoir high in the hills, the flume was a man-made river that ran for 20, 40, even 50 miles, clinging to the walls of canyons, bridging valleys like a barbarian's version of a Roman aqueduct, and carrying down from the heights not only saw logs but lumber, shingle bolts, cordwood, posts and railroad crossties at speeds up to a mile a minute. To achieve its primary purpose — transporting sawed products — it depended on a small sawmill near the logging site to cut the larger logs into planks, which could then be fed to a terminus below: a more sophisticated mill, a drivable stream, or a logging railroad that would carry the timber to market.

The idea was born in Nevada, where one J. W. Haines had successfully floated sawed timbers from a mill on the east side of the Sierra down to the Carson Valley for use in the mines of the Comstock Lode after the great silver strikes of 1859. Haines' flume was 12 miles long, but it was a mere midget when compared with some of the incredible contraptions that came into use later.

In the early 1870s Clement F. Ellsworth, a young logger from Maine, constructed the first major flume in the California forests. From his Belle Sawmill on the edge of a mountain meadow near Lyman Springs, the flume carried lumber 40 miles to the town of Sesma on the Sacramento River. It was three years in the building, and for most of that time was referred to as Ellsworth's Folly. People said that if the flume ever was completed, the boards that it carried would be as costly as gold dust, and they could hardly be blamed for thinking so.

Ellsworth designed his flume so that in cross-section it was 48 inches wide at the top and 16 inches wide at the bottom, with 32-inch sloping sides. During construction, the flume and its supporting trestlework needed about 135,000 board feet per mile of Ellsworth's lumber and, like the many flumes that were patterned after it, the structure was its mill's own best customer while being assembled. The flume itself was prefabricated at Ellsworth's Belle Sawmill in 16-foot sections called boxes, with an intended grade of one inch per section, or 27 feet per mile—for a total fall of 1,080 feet. But a constant gradient was impossible to maintain; in some places water in the flume ran as fast as 50 miles an hour, and in others—on flat valley crossings—it barely flowed at all.

Every few miles along the route a cabin was built to house a flume tender, or lumber herder. His job was to patrol a narrow catwalk running alongside the flume looking for jams or leaks, the bane of flume operators. A load of lumber would pile up on a curve and the weight could cause a flume section to collapse, allowing all the water and lumber coming from above to spill down on the hillside below. When lumber was being moved at night, a tin can was hung in the flume near each herder's cabin. If the can stopped rattling as boards struck it, the herder was supposed to wake up and prepare to cope with a problem above his post. In the dark, he would pick his way upflume along the catwalk in search of the jam or break. In rainy, snowy or icy weather the eight-inch catwalk was a precarious walkway, and from time to time a lumber herder slid off into the deep darkness.

In fact, the catwalk was poor Clement Ellsworth's own undoing: one day in 1873, while he was supervising the final phase of construction, he tumbled from the catwalk, "suffering a very severe jar to his whole system and seriously crippling his feet," as a Sacramento Valley weekly paper reported. Actually, Ellsworth's injuries were fatal and he died too soon to see his "folly" vindicated. The flume reached Sesma in August 1873, shortly after his death; and for three years thereafter, with the exception of occasional pauses for accidents, floated at least 40,000 feet of lumber

In a boat specially built for the adventure, two thrill-seekers hurtle down a 15-mile flume near Carson City, Nevada. *Harper's Weekly,* which commissioned the drawing, said they made it in less than 11 minutes.

daily down from the mountains to supply a planing and moulding mill in Sesma.

In addition to transporting lumber, flumes were the waterways for crude little boats used to ferry all sorts of articles—from crates of groceries to catches of fish and game for friends in town. Once in a while an injured logger was floated down the flume to a doctor; and some people rode them just for the excitement of it. The flume boat varied in size and shape depending on the needs of the moment, but it usually had a V-shaped keel to conform with the shape of the flume and a plank platform running from stem to stern to serve as a deck. Packages could be lashed to the platform—or could be held by passengers who were bold enough to hazard the ride.

Such a trip was always dangerous. The boat could take a curve too fast and leap right out of the trough, or

103

Logs roar down from an Oregon splash dam seconds after the barriers are released. The function of these sturdy reusable dams was to accumulate enough water to move felled timber down shallow streams.

a large splinter projecting from the sides of the trough could impale a passenger. Both kinds of accidents happened regularly.

In 1903, a journalist by the name of Bailey Millard ventured down a flume even bigger than Clement Ellsworth's and lived to describe his adventure for *Everybody's* magazine. The Sanger Flume was the longest and steepest in the world. It ran 54 miles from Millwood in the heart of the Bigtree country to Sanger in the San Joaquin Valley. The descent was 4,737 feet all told, 4,300 feet of it occurring in the first 13 miles — a precipitous drop of 330 feet per mile. A pair of San Francisco lumbermen-promoters, Hiram C. Smith and Austin D. Moore, had spent four years building the flume. It was finally completed in 1890 at a reported cost of $300,000, and had used nine million feet of clear, knot-free redwood lumber in the process. The two San Franciscans operated the flume for five years and then sold out to the Sanger Lumber Company, which was floating out 250,000 board feet of redwood lumber daily by the time Millard took his eventful ride.

As Millard and his photographer clambered into the makeshift boat, Frank Boole, the company manager, told them: "I wouldn't ride down that flume for the whole plant." And a flume hand added as he shoved them on their way: "You'll want to hold on. She'll run like the milltails of hell."

The two men took their "seats," the photographer on a redwood block and Millard on a cracker box. A few minutes later, on a hell-for-leather stretch called the Devil's Slide, Millard shrank down off his perch and clung to the crude deck in utter panic while "the whizzing landscape fused into a long, filmy, biographic blur. I could not time the swiftness of the Devil's Slide any more than a man riding a cyclone can manage a stopwatch, but it seemed to me that no express train ever equalled it."

Farther along on the ride, the boat tobogganed under "high, tragic cliffs on the edge of which the flume clung by a sort of miracle in which even the gnarled pines had come to have faith. The speed increased and we swam dizzily out upon a terrible trestle, the spindling timbers of which, as seen from an oncoming curve, seemed the flimsiest of supports. We crossed one creek thirteen times, darting out upon the bridges unexpect-edly from curves that nearly swayed one to perdition.

"After passing the awe-inspiring Dingwald's Trestle and the Shotgun Trestle, we soared over Davidson's Flat leaning out from heaven to see where a grub-boat had dashed through the flume on a certain fateful flight. There lay the splintered fragments far below, and the wreck of a cook-stove which had gone down with it — a scrappy pile of junk which not even the covetous Indians had seen fit to carry away."

Along the 54-mile run, he commented, there was "enough waste lumber to build a good-sized town. The planks and stringers, floated down in long trains from the mill to the valley below, wedge together in great jams, and the lumber that follows is diverted by the rising water thus banked up in the flume and is thrown down into the cañon."

On one curve Millard ducked aside to dodge an overhanging oak limb, lost his balance and was thrown out of "the crazy little craft." Fortunately he fell backwards into the trough itself, where he floundered around in the icy water listening to the diminishing sound of unsympathetic laughter from his companion. Drenched and bruised, he ran to catch up with the craft, managed to do so on a slow stretch of the flume and scrambled safely back aboard.

"The ride," wrote Bailey Millard upon returning home after his exhilarating experience, "is such a bit of brisk living as sets the blood all a-tingle and gives one a taste of the recklessness of Phaëton trying to drive the chariot of the sun. One feels that to make such a voyage every day would in time fill even the commonest of men with the abandon of the gods."

The ideal flume, of course — one capable of floating logs and timbers of all sizes — was a river. Natural waterways had been the conventional and proper way to transport timber out of the woods in the Lake States and in New England. Out in the Far West, the loggers mounted great drives, with huge, Maine-style rafts of logs, down the major rivers flowing to the Pacific Ocean: the Columbia, the Umpqua, Rogue, Klamath, Eel and Sacramento. In the 1890s, when lumbermen started invading the Rockies and the vast pine forests of Idaho, loggers also mounted traditional white-water drives on the St. Maries and St. Joe, which raced down from the mountains to Lake Coeur d'Alene. ◉

Dolbeer's donkeys: the little engines that could

The donkey that finally ousted oxen from the skidroads of the West was not a beast but a small steam engine. It was called a donkey, after a ship's auxiliary engine, because the man who invented it in 1881, a mechanical genius named John Dolbeer, had been a naval engineer before turning to logging.

Dolbeer's inspiration was to use a small, high-pressure steam engine to turn, through sets of gears, a capstan-like spool or drum that could reel in rope with the power of many horses, mules or oxen. Lashed in place, the engine could reel in the heaviest of logs. At first it was used to "yard," or haul, logs from where they fell just a few hundred feet to the head of a skidroad, where the bull teams took over. Within a few years, however, after wire cable replaced manila rope, loggers learned to "road" the logs hundreds upon hundreds of yards from one donkey to another along the entire length of the skidroad. And that was the end of the bulls.

Donkey engines kept getting bigger and more versatile. There were two-drum and three-drum donkeys able to haul in logs from the woods over high wires, donkeys that loaded logs on railroad cars, and landing donkeys like the one at right that pulled logs to river or lake landings. In 1902 the donkey even became a steamboat, as at upper right. On California's Big River, one lumber company launched a converted lighter with a donkey engine that turned a stern wheel; christened S.S. *Maru,* it was used to herd large "rafts" of logs downriver. The donkey remained the power of the forests until the internal-combustion engine did to the donkey what the donkey had done to the bulls.

Lashed to a heavy log foundation at the edge of a river, a landing donkey snakes a log toward the water shortly after the turn of the century. Logs already launched are gathered in a raft to be floated to the mill.

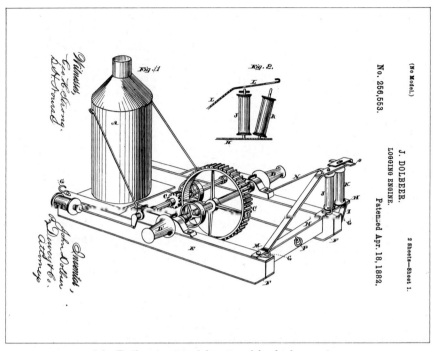

John Dolbeer's original drawing of the donkey engine.

S.S. *Maru,* a converted stern-wheeler powered by a smoke-belching donkey, chugs along the Big River. The second smoke plume issues from the cookstove that was used to prepare hot meals for the crew.

A redwood log yarded by a Dolbeer donkey engine in the 1890s stops just short of the pulley through which the cable passes. The stump at center is one of several to which the engine was lashed for stability.

Anchored on a steep Oregon hillside in the 1890s, a donkey reels in a cable hauling a log out of the woods. The only man working at this point is the spool tender, who controls the pulling power of the winch.

A rail-riding donkey hoists into place a section of a trestle over Tornado Creek in California about 1910. Logging railroads had to cross increasingly rougher terrain as operations moved farther into the mountains.

The Idaho lumberjacks would cut all winter, stack their logs on the banks of the frozen rivers, and then shoot them downstream to the lake when freshets came in the spring. But like almost everything out West the shooting was outsized, surpassing anything on the Penobscot or the Kennebec in Maine. The St. Maries River curled like a snake through 60 miles of wild country to meet the St. Joe, an equally twisting torrent of ice water that tumbled westward out of the rugged Bitter Root range. Below the confluence the merged stream was amiable and navigable — at 2,200 feet above sea level, the highest navigable river in the world — and in the final 30-mile stretch down to the lake, "boom men" could at their leisure sort out logs stamped on the ends with the marks of dozens of upstream loggers.

But on the higher reaches were such treacherous rapids as The Loops on the St. Maries River, where for nearly four miles the water was white and deafening, beautiful and frightening. Both of the narrow rivers were prone to monumental jams. All it took was a single log, wedged across the stream. As more logs piled up behind it, the dammed-up water rose, and the logs in the rear of the drive spilled out onto the river banks. It could take days or even weeks to find and pry or blast loose the "key logs" — and then there were only seconds to get out of the way, to escape being crushed by a pine trunk or drowned under the rush of the released water and wood.

At times, coastal rivers, such as the Coquille and Coos in Oregon or the Queets, Humptulips and Wishkah in Washington, were used for a different kind of log drive, one that employed a deliberately created torrent of water in the shallow streams. The loggers built a splash dam of timbers where a river squeezed between high banks, and stored their harvest of felled trees in the pond in back of the dam. When the pond was swollen to capacity, the spill gate in the splash dam was pulled away, releasing a flood that the loggers called the splash. The rush of water sluiced the stored logs — and any others that had accumulated along the watercourse — down the stream toward the sawmills. At one time on the Humptulips River there were as many as eight splash dams in operation. They had their disadvantages; after "splashing the dam," many logs were hung up along the river bank below. Too

massive to be rolled into the dwindled streams by muscle power, they remained stranded until a bigger splash came — or waited there for years until new technologies made it possible to salvage them.

Over the decades, the strenuous, chancy chore of getting the logs out had been undergoing a slow process of evolution. At last, in the 1880s, technology transformed logging dramatically and forever. Instead of relying on man and animal power, or the forces of gravity and water, inventive geniuses dreamed up a family of strong, steam-driven machines to help with the work. And as they appeared at the logging sites in increasing numbers, logging was revolutionized. An era of what lumbermen called highballing began: a breath-taking acceleration in the movement of logs and lumber. And bedlam shattered the quiet of the woods.

Steam power, by itself, was nothing new in the lumber business. Steam sawmills had been cutting wood for 40 years, almost from the beginning of the industry. Yet only in the early 1880s were lumbermen able to adapt the steam engine to the needs of the crews back in the tall timber.

For years, bearded Ephraim Shay of Haring, Michigan, had been seeking ways to speed up the leisurely movement of logs from the forest and his mill. He had tried using horses to drag log-carrying cars running on maple rails, set far enough apart so that a team could plod between them. This crude railroad was an advance over the team and the skidroad but, as Shay said, "the cars would catch the horses on downgrade and sometimes kill them." Not satisfied, he kept searching for a better idea.

Shay had seen fellow lumbermen use railroad locomotives on iron-faced tracks to pull trains of logs. But their effectiveness was sorely limited in mountainous timber country. They were often too heavy for the loggers' wooden trestles and poorly ballasted roadbeds. Furthermore, the powerful thrusting rods that ran from the pistons to the driving wheels provided uneven traction; the wheels would spin under heavy loads and, on grades greater than 1 per cent (one foot of rise in 100 feet of track), the engines lost traction altogether. Attempting to overcome the problem, some lumbermen commissioned lightweight engines with double-flanged wheels that ran on "pole roads" of wooden

A Shay locomotive, drawing a log train, negotiates a curve in Washington. Driven by gears instead of rods, it maintained excellent traction and power on steep, winding grades where conventional engines faltered.

rails. They were not much of an advance. But for Shay, the experiments contained the germ of an idea.

All through the 1870s Shay tinkered with the design of a radical engine that would be light in weight and that would work without the conventional piston-and-rod arrangement. Ephraim Shay's better idea was introduced to the logging world in 1880 and it looked like some mad inventor's nightmare. "The pilot model of the locomotive that was destined to become respected around the world," wrote logging historian Kramer Adams, "consisted of a short railroad flatcar with a wooden water tank at one end, a wood bin at the other and an unsightly assortment of machinery surrounding an upright boiler in between. What's more, the thing was lopsided, with the boiler on one side, and geared transmission machinery on the other."

Unsightly it may have been. But it effectively incorporated Shay's central idea: power from the lone cylinder was transmitted by gears, not rods, to the four driving wheels in each of two trucks mounted on a single chassis—eight driving wheels in all. And the gears were the key to greater and steadier traction. A geared shaft ran from the engine to each wheel, meshing with a ring of gears on the wheel. Locked in its gears, the wheel could not spin. The Shay locomotive was an all-weather, all-terrain log mover.

"My friends remonstrated with me," Shay remembered later, "for spending so much time and money on such a crazy idea, and in fact, they really thought I was a little cracked, and did not hesitate to say so."

By June 14, 1881, when Shay was awarded Patent No. 242,992, he was already at work on more powerful two- and three-cylinder versions of his invention. In time, thousands of Shay-geared engines were built by various locomotive works; and while many served in the Lake States, where the inventor himself logged, most of them labored in the much steeper forests of the West.

The Shay was widely admired. It won a gold medal for excellence in 1905 at the Lewis and Clark Exposition in Portland, and loggers swore by it. They contended that a Shay could hunker down and climb a tree if it had a mind to, and that was only a slight exaggeration. Its smoothly meshing gears enabled it to haul cars full of logs up grades as steep as 3 per cent, and its relatively light weight in proportion to its pow-

er was a crucial factor on flimsy logging roadbeds. Among the locomotive's features were universal joints in the undercarriage that enabled the wheel-trucks to swivel independently; they made the Shay look disjointed when it came around a curve, but they enabled the vehicle to hold to the rails on a bend so extreme —so the loggers insisted—that the headlight would shine right back over the engineer's shoulder into the firebox behind him.

Ephraim Shay never got rich, or wanted to, from his invention. The first engine he sold brought him exactly $1,070. When he turned his design over to other manufacturers, he accepted $10,000 in royalties and then refused further payments. Shay was evidently content to know that, from 1880 on, the fame of his name spread with every geared locomotive that chuffed into the deep forests.

One day in August 1881, two months after Shay received his patent, a mechanical monster smaller than any locomotive but bulkier than a yoke of oxen appeared in the redwoods back of Eureka, California. It sat on heavy wooden skids and consisted of an upright wood-burning boiler with a stovepipe on top, and a one-cylinder engine that drove a revolving horizontal drive-shaft with a capstan-like spool at each end for winding rope. It was the first Dolbeer Steam Logging Machine, and it would be recorded as the greatest labor saver in the industry's history.

The inventor of this "donkey engine"—so called because early models, though mighty, looked too puny to the loggers to be dignified with a rating in horsepower —was a successful logger named John Dolbeer, of the Dolbeer & Carson Lumber Company. He had gone West in 1850 as a 23-year-old from Epsom, New Hampshire, to join the gold rush. William Carson, his future partner, was an adventurer of 25 from New Brunswick, Canada, and he also was after gold. Along with most gold seekers the two Easterners were in for disappointment; but unlike most they were lucky enough to have a trade to fall back on, having worked in the Northeast woods. Dolbeer headed for Gold Bluffs, near Redwood Creek in Humboldt County; Carson went inland to the rugged Trinity River region.

Neither of them struck it rich immediately but both made sufficient money by 1853—Carson as a contract logger, Dolbeer as a millworker — to buy interests in small sawmills and eventually become millowners in their own right. In 1864, the two decided to join forces to operate Dolbeer's sawmill at Eureka in Humboldt County. So successful was their partnership that the Dolbeer & Carson company name — the equivalent in redwood country of Pope & Talbot — endured for more than 80 years.

Carson was the logging boss and mill operator; Dolbeer's contribution to the partnership was his indefatigable development of problem-solving devices. Even before Carson joined him, Dolbeer had replaced his mill's old up-and-down gang saw with a faster circular saw of his own invention. In time, Dolbeer replaced that saw in turn with a double circular saw; and then, forever modernizing, he equipped his mill with the first band saw in the West: an endless, razor-toothed steel belt that could cut at high speed through a log of any size with minimal waste *(page 115)*.

Dolbeer visualized the donkey engine as an infinitely more efficient way than ox-power to snatch logs out of ravines and haul them to the head of a skidroad or to the side of a stream. If steam could power a pile driver, help to load and unload ship cargoes and drive a sawmill or a locomotive, Dolbeer figured that it surely could be adapted for work in the woods.

On Dolbeer's maiden model, patented in April 1882, a manila rope 150 feet long and four and a half inches in diameter was wrapped several times around a gypsy head: a revolving metal spool mounted on a horizontal shaft. The loose end was carried out to a log and attached to it. Then, with a head of steam built up in the high-pressure boiler, the spool revolved and the incoming rope hauled the log toward the donkey.

Operating an early Dolbeer donkey required the services of three men, a boy and a horse. One man, the "choker setter," attached the line to a log; an engineer, or "donkey puncher," tended the steam engine; and a "spool tender" guided the whirring line over the spool with a short stick. (An occasional neophyte tried using his foot instead of a stick; when he was back from the hospital, he would use his new wooden leg instead.) The boy, called a whistle punk, manned a communicating wire running from the choker setter's position out among the logs to a steam whistle on the donkey engine. When the choker setter had secured the log to

the line running from the spool, the whistle punk tugged his whistle wire as a signal to the engineer that the log was ready to be hauled in. As soon as one log was in, or "yarded," it was detached from the line; then the horse hauled the line back from the donkey engine to the waiting choker setter and the next log.

For years the job of hauling back the line kept hundreds of "line horses" employed, until a simple improvement put the horse out of work: a "haulback line" was joined to the main line, and the two were converted into a long loop, the far end of which traveled through a pulley block anchored in a tree stump. Out among the felled trees the choker setter fastened a noose of cable around a log and attached it to the main line; the noose "choked" the log securely when the line from the donkey engine grew taut and began to pull. When the log reached the yard and was released, the haulback line pulled the main line over the spool and dragged it back into the timber.

Over the next decades donkey engines evolved far beyond the stage of Dolbeer's initial fragile invention. Donkeys were mounted on barges to herd rafts of logs, "road donkeys" pulled logs along skidroads, and "bull donkeys" lowered entire trains of log cars down steep inclines *(page 111)*. All of these descendants took advantage of another sophisticated invention: the wire cable, first made of iron and later of steel.

Metal cable had been available since shortly before the Civil War, but Dolbeer did not use it with his early donkey engines because it kinked and often snapped. But manila ropes, especially when wet, stretched and were too hard to handle for hauls of more than 200 feet. By the 1890s strong steel cable, winding and unwinding on rapidly turning drums, gave the donkey engine a pulling range of 1,500 feet, along with the strength to do just about anything that needed doing in the woods. Like the Shay locomotive, the Dolbeer donkey inspired many imitators and variations, some sensible and some hare-brained. Apparitions like the Lidgerwood "skidder" — a powerful, donkey-type engine mounted on a heavily built railroad car — showed up in the woods to haul logs. There was a machine called a Walking Dudley, which towed a turn of logs between tracks to a yard, then "walked" back for more, hauling itself on tracks by means of a steel cable that it reeled onto a huge wheel in its midsection. ◉

The everlasting quest for the perfect saw

Over the years the power-driven saw evolved to satisfy one goal: to cut bigger logs faster, and with less waste. In the early days, the standard "head rig" — the first major cutting device a log encountered at the mill — was a gang saw (*right*) composed of long, narrow blades that used an up-and-down reciprocating action. The blades were a fixed distance apart and sliced up every log identically, regardless of flaws. The gang saw was woefully inefficient because the engine wasted energy overcoming inertia on every stroke and return.

Starting around 1860, these saws were gradually replaced by faster, more efficient circular saws. But the depth of a circular saw's cut was necessarily limited to its radius. By the 1870s, the double circular saw — with one blade set above another — was sawing bigger logs into planks of any desired thickness by making simultaneous cuts in a log carried back and forth on an adjustable carriage (*below*).

A subsequent development raised the size limit of the double saw by adding a "breaking-down saw." This contraption (*opposite*) reduced outsized logs to manageable size by means of an additional vertical saw, set a few inches farther out over the log, plus a small horizontal circular blade. Together they lopped boards off the log's top, "breaking it down" to a size that could be handled by a double saw.

The breaking-down saw was a California invention, intended for huge redwoods. But the redwood challenge was not really met until the adoption of the band saw in the 1880s. A huge loop of saw-toothed steel that was stretched between flywheels set above and below the carriage, this epoch-making saw was able to reduce to planking the biggest trunks of the redwood forests of the West Coast — or even of the Sierra.

A log rides the carriage of an early gang saw: six vertical blades a fixed distance apart and geared to a pair of steam-powered flywheels.

Three operators stand by their double circular saw at a mill in Oregon country (*above*). Their functions are delineated in the 1870s engraving at the left: two loaders roll a log toward the carriage, where the sawyer waits to lock it down so that it can shuttle back and forth through the blades.

The sawyer *(left)* and the crew of the Hobbs Wall Mill at Elk River, California, show off their breaking-down saw in 1895. This intricate, four-bladed saw could handle the biggest redwood, but it was soon superseded by the more efficient band saw.

Between cuts, the sawyer pauses by the band saw of the Union Lumber Company in Fort Bragg, California, in 1900. The offbearer, who removed the cut boards, is standing in front of the previous slab. The flywheel arrangement is detailed in the engraving *(above)* of a smaller saw.

But none of the ingenious improvisations ever was as widely favored as John Dolbeer's snorting donkey. The engine ushered in the era of what was called ground-lead logging, in which the incoming log was hauled full-length over the ground by the donkey engine's cable, bumping and battering everything that stood in its path. A big "donkey show," as the logging operation came to be called, was usually a hair-raising spectacle. A turn-of-the-century author, Ralph D. Paine, happened upon just such a scene in the Western Cascades and was filled with both admiration and terror. "Stout guy ropes ran to nearby trees, mooring the 'donkey' as if it were an unruly kind of beast," he wrote. "In front of the engine was a series of drums, wound round with wire cable which trailed off into the forest and vanished. The area was littered with windfalls, tall butts, sawed-off tops and branches, upturned roots 15 feet in the air. Huge logs loomed amid this woodland wreckage like the backs of a school of whales in a sea."

Paine noticed a long signal wire that led away from the engine's whistle off into the woods. "Someone out of vision yanked this. The 'donkey' screamed a series of intelligent blasts. The engine clattered, the drums began to revolve and the wire cable which seemed to wind off to nowhere in particular grew taut. The 'donkey' surged against its moorings; its massive spread began to rear and pitch as if striving to bury its nose in the earth.

"There was a startling uproar in the forest, wholly beyond seeing distance, mind you. It sounded as if trees were being pulled up by the roots. In a moment a log came hurtling out of the underbrush nearly 1,000 feet away. It burst into sight as if it had wings, smashing and tearing its own pathway . . . so fast that when it fetched athwart a stump it pitched over it as if it were taking a hurdle. Then it became entangled with another whopper of a log, one as big as itself. The two locked arms, they did not even hesitate, and both came lunging.

"It is an awesome sight to see a log six feet through and 40 feet long bounding toward you as if the devil were in it, breaking off some trees as if they were twigs, leaping over obstacles, gouging a way for itself." When the gigantic log was within 20 feet of the loading platform where he stood, Paine panicked and ran,

but then "the huge missile halted in its flight, and the masterful donkey had a breathing spell."

At a different yard at the end of that same day, Paine watched another donkey at work, perched on a logged-off hilltop. As the engine pulled a faraway log homeward, its wire cable led down the hill, across a pond and up a furrowed trail into the forest. Suddenly the log charged out of the woods, hurling earth and stones before it.

"On top of it stood a logger, swaying easily, shifting his footing to meet the plunges of the great beast, a daredevil figure of a man outlined against the sunset sky as the log flew down the hill," Paine wrote, "Before it dived into the pond he made a flying leap and tumbled into the underbrush with a yell of pure enjoyment." Then the log tore through the pond in a whirlwind of spray and climbed the near slope to the yard where the donkey released it.

The donkey engine made possible an even more awesome and exciting kind of logging than the spectacle witnessed by Paine. It was called high-lead logging and it required the services of a nerveless daredevil known as a high climber, who had to be part woodsman and part steeplejack. The occupation was born by accident; some logger, watching a cable whir through a pulley set high on a tree trunk as it yanked a log toward a donkey, got to wondering: why not put the pulley clear on top of the tree, the better to haul logs over obstructions?

The high climber's job was to prepare such a "spar tree." Equipped with saw and ax, spurs and rope, he scaled the tree, lopping off limbs on his way up. At a point 25 to 30 feet from the top and 150 to 200 feet in the air, he would dig in his spurs and belt himself to the tree with his rope. Hanging precariously, he then topped the tree, chopping or sawing off its crown or perhaps setting dynamite charges to blast off the topmost 25 feet or so.

If the man was a "high rigger" as well as a high climber, he could climb back up the decapitated tree and attach a heavy pulley block on the top of it, while guy lines were affixed to the spar to hold it steady against the tremendous force that would be exerted by several tons of plunging logs. When completely rigged, the stout spar tree supported a main line or lead block weighing a ton or more, and through the block ran the cable that would be connected to the enormous logs lying on the forest floor.

When the whistle punk gave the go-ahead signal, the engineer opened the donkey's throttle and the line sped through the block on top of the spar. Out in the forest, the incoming log suddenly reared up like some berserk prehistoric monster and went charging, one end high in the air, through the brush — until the donkey engine dropped it with a resounding crash among its companions in the yard.

The amazing donkeys had a profound effect on the economics of logging. Simon Benson, a lumber magnate who worked the forests along the Columbia River, found in 1899 that his costs in getting the logs out dropped from $4.50 to $2.10 per thousand board feet when he switched from ox teams to steam donkeys. And, as a lumberman named William Kyle succinctly put it, "When the machine don't work it don't cost anything to keep it, and you don't have to feed it when it is not earning anything."

Though in time the donkey eliminated most of the horses and bulls, it gave jobs to many men. Before the coming of John Dolbeer's spark-spewing marvel, a score or so of men might make up a logging crew; afterward, with production highballing along, a camp of 200 men or more was not unusual.

But if there were jobs for many men, there were not jobs for all. The bull whacker was doomed, like his beasts, to banishment from the woods, and with his departure a picturesque era in logging came to an end. One night, sitting idle in a bunkhouse, Dan McNeil, who had been a noted bull whacker in his day, composed a requiem for his trade:

> *Then I was king of the whole woods-crew,*
> *   and I ruled with an iron grip;*
> *And never a slob on the whole dam' job*
> *   dared give me any lip.*
> *But now, alas, my days are past;*
> *   there's no job for me here.*
> *My bulls are killed and my place is filled*
> *   by a donkey engineer.*
> *Instead of my stately team of bulls*
> *   all stepping along so fine,*
> *A greasy old engine toots and coughs*
> *   and hauls in the turn with a line.*

117

River pigs, bull whackers and other timber beasts

Though a host of labor-saving mechanical marvels appeared in the woods during the 1880s, it still took the sweat and sinews of all sorts of loggers — or timber beasts, as they were often called — to topple the tall timber of the West and deliver it to the sawmills.

Manpower was indispensable for doing many jobs, from the highly skilled fallers to the unsung peelers and buckers who transformed felled trees into uniform logs. It took cool-nerved choker setters to hook the logs by cable to donkey engines, and nimble skid greasers to ease the way along skidroads. The ox teams were driven by bull whackers whose profanity, legend has it, could steam bark off a tree. And similarly, the donkey engineers developed their own blistering vocabulary as they ran their temperamental machines.

A breed apart were the river pigs who mounted river drives. Each day a crew might herd some 10 million feet of logs three or four miles downstream while breaking up jams and refloating stranded logs. Fittingly, a river pig was one of the highest-paid woodsmen. As an oldtimer said: "He got more money because he was wet all the time."

FALLERS

Perched on springboard scaffolds, a pair of fallers tackles
a redwood. The men have chopped through part of the trunk
with axes, now lodged in the tree below the cut at center.
To deepen the undercut, they have chosen a long saw and
driven wedges into the cut to keep the blade from binding.

PEELERS

Members of a peeler crew, busily stripping bark from a
felled California redwood, form an unusual backdrop for a
photographer's portrait. The sharp-edged peeling bars were
used to remove the redwoods' heavy, often foot-thick
covering—a procedure rarely needed for thin-skinned firs.

BUCKERS

After neatly slicing a felled and peeled redwood log, buckers
split the timber into "shingle bolts" — short segments to
be shaved into redwood shingles at the mill. When the
tree was to be reduced to boards, the buckers cut the trunk
into designated lengths, usually of 16, 24, 32 or 40 feet.

CHOKER SETTERS

A crew of choker setters, having secured and cinched a
steel cable around a fir, waits for a donkey engine to haul
the log out of the woods. Though dangerous, their job
required little skill; as a result, choker setters were among
the lowest ranking and poorest paid of a forest crew.

BULL WHACKER

His goadstick ready, a bull whacker leads a yoke of oxen
onto a skidroad in Mendocino County, California. Bull
whackers used as many as 10 yokes to drag a string of
logs, urging on the beasts by cursing, prodding and, when
all else failed, walking on their backs in calked boots.

SKID GREASER

A youthful skid greaser prepares to set off down a skidroad
with his tools: a pail of grease, and a swab and broom to
spread the lubricant over the skids. Greasers used whatever
was available: rendered dogfish oil or animal fat, tubs of
rancid butter or, if nothing else was at hand, ordinary water.

FLUME HERDERS

Spread out along a stretch of flume in the Sierra, flume
herders refasten sawed planks into bundles after they were
jarred loose during the downhill ride. The men in the
foreground are using pickaroons to intercept the loose
planks. Flume herders were also responsible for clearing
jams and for making sure that the flume was in good repair.

CHUTE FLAGMAN

Seated in his crude shanty near the base of Oregon's
Pokegama Chute, an elderly flagman—undistracted by
visiting friends—keeps a sharp eye on logs hurtling past at
speeds averaging 90 miles per hour. The flag was a signal
to the loaders uphill that the chute was clear. During
the night, a lighted lantern served the same purpose.

RIVER PIGS

Logs stranded along the Willamette River are muscled into
deeper water during a river drive. The rollers *(rear)*
extricated logs with peaveys; teamsters *(foreground)* freed
logs aground by hitching their horses to "dogs"— metal
spikes driven into log ends by the maul-wielding doggers.

BOLT PUNCHER

On Washington's Skagit River, a bolt puncher—or pond
monkey—uses a pike pole to ease a shingle bolt onto
the conveyor belt that led to the mill. Bolts were kept in
ponds as an easy method of storage and handling, and
to be sure they were clean when they went into the saws.

4 | A rude and perilous life

At a typical 1890s logging camp the bunkhouse was the nearest thing to a social center. Inside, of an evening or a rainy Sunday, lumberjacks sat or lay about, washed clothes, played cards, complained about the "gut robber"—or cook—argued, shaved if they were ambitious, or tried to read. Mostly they talked—spinning tall tales about Paul Bunyan, their outsized mythical hero, and all-too-real stories about comrades lost to falling timbers, flailing machinery and the forest fires that were their daily hazard. The bunkhouses were crowded, dark, and redolent with a rich mixture of odors: drying woolen shirts and socks, wood burning in the cast-iron stove, sweat and snuff and kerosene, the sodden contents of the cedar-chip-filled spittoon and, more often than not, exhalations of whiskey.

Those were the standard smells. One logger reminisced about his first day, while he was young and green, at camp: "I reported to the foreman and asked where I should go. He pointed to the bunkhouse and told me I should have to build my own bunk. He told me I should find lumber down at the hog lot. I built something like a bunk, and then went down to the barn to rustle straw to fill it. Over this I spread blankets. That night the hard boards and the odor of hogs kept me awake for a long time."

If his bunk didn't keep a logger awake, the lice were likely to. Even the methods used to get rid of them contributed to his discomfort. At Wendling, Oregon, exterminators sealed the bunkhouse windows and ran the sawmill steam pipe in through the door. After pumping the place full of steam, they looked in to find a few stray vermin crawling irresolutely out of the walls to die—and the loggers' crude furnishings warped into unusable shape.

On their own time, loggers crowd a Washington bunkhouse in 1892, drying shirts and socks

(and feet) by the heat of the wood-burning stove and playing cards by the light of a kerosene lamp. On Sunday many went out hunting deer.

The heroic legends of hard-driving men

In the harsh, often cruel world of the pioneer loggers a man needed as potent a talisman as he could devise. It gave the timber beasts and river pigs the will to mock their all-too-obvious inadequacies, and the borrowed courage to cope with the thousand dark anxieties that beset them — from fear of the potential destruction locked in a log jam to fear of being maimed by a hurtling tree or a whipping cable and left unemployed, hungry, superfluous.

The men of the American woods fabricated a marvelous charm. They invested it with superhuman qualities and called their creation Paul Bunyan. No one knows how they hit on the name, for the origins of his legend are lost in the cool recesses of the vast continental forests. But he lived in the minds and hearts of woodsmen everywhere as surely as if he stood beside them, and shared their lives and witnessed their deaths. The lumberjacks spun their tales of Paul Bunyan at night by the flickering light of bunkhouse lanterns, and passed them from camp to camp by word of mouth. Over the years, the myth grew: Paul Bunyan became the essence of the mighty logger — huge and heroic, bigger than Goliath, stronger than Hercules, more fearless than Spartacus. And if Paul Bunyan could roar with laughter at danger and perform wondrous feats, then some of his disdain for the hazards of the forest was bound to rub off on those who believed.

An old logger, around the turn of the century, recounted the stirring story of Paul's infancy: "A lot of people wondered where Paul Bunyan was born. But he was born in Maine. When he was three weeks old, he was such a lummox of a kid that he wallered around

A trail crew opens a road blocked by the debris of a fire in August 1910. The holocaust destroyed two billion board feet of wood in northern Idaho and Montana.

so much in his sleep that he rolled down four square miles of standin' timber. Well, the natives wouldn't stand for that, so they built him a floatin' cradle and anchored it out at Eastport, Maine. Every time he rocked in that cradle, he caused a seventy-five foot tide in the Bay of Fundy.

"And when he got asleep, they couldn't wake him however, so they called the British navy out and a-fired broadsides for seven hours. When they did awake him, he was so excited over so much excitement, that he tumbled overboard into the ocean. And he raised the water so it sunk seven warships. Well the natives wouldn't stand for that, so they captured his cradle, and out of the cradle they made seven more ships. But the tide, in the Bay of Fundy, is a-goin' yet."

To hear the lumberjacks tell it, Paul Bunyan never did anything small. As a young man he lived in a great cave, and one winter night he was awakened by a loud crash outside. A falling body had shattered the ice on a nearby frozen lake, fracturing it into tossing, seven-foot-thick floes. Far out in the lake Bunyan spied a pair of ears protruding from the water. He waded a mile out — up to his chest — grabbed the ears of a baby calf and dragged it ashore. It was blue all over, either from the cold water or because that was The Winter of Blue Snow, which turned everything blue. The calf stayed blue even after Bunyan nursed it to health, named it Babe and made it his boon companion.

Loggers never did agree on the exact size of the fully grown blue ox; the width between its eyes, for example, measured either seven ax handles or 42 ax handles (plus, to be precise, a plug of chewing tobacco). In any case, Babe was big enough to pull anything with two ends to it and thus help Paul Bunyan and his crew log the country from one end to the other.

Nobody ever took Bunyan's measure either, but it was said that whenever he ordered a new pair of boots they were delivered lashed down on two railroad flat-cars. He had a fine head of curly black hair, which his wife lovingly combed each morning with a crosscut saw after parting it with a broadax. And he took great pride in his beard, several times daily pulling up a pine tree by the roots and using its stiff branches to keep the hair smooth and shiny.

Naturally, Paul was the fellow who invented all the woodsman's tools, all the techniques, and a lot of other things in addition. He got the idea for the grindstone by rolling a huge boulder downhill and running along-side it, holding his ax flush against the whirling granite. While out felling trees, Paul liked to put his mark on the trunks by pinching pieces out of them with his fingers. When he had marked enough of them in this manner, he would start to work with his four-bladed ax, which he would swing around his head, chopping down four trees at a time. On days when he felt really ambitious, Paul would haul out his great timber scythe and mow down whole sections of forest at a single sweep—although occasionally he preferred to use his three-mile-long crosscut saw.

With Paul and Babe around, lumberjacks never had to worry about getting their logs out of the woods to market. An oldtimer once laboriously wrote down the bunkhouse tale of how the greatest logger and his blue ox dealt with a Lake States log jam:

"Paul B Driving a large Bunch of logs Down the Winnonsin River when the logs suddenly jammed in the Dells. The logs were piled Two Hundred feet high at the head, And were backed up for One mile up river. Paul was at the rear of the Jam with the Blue Oxen And while he was coming to the front the Crew was trying to break the jam but they couldent Budge

An ox team stands beside the logs it has pulled down a skidroad to a logging camp railhead. Spare animals are penned in the corral at center.

it. When Paul arrived at the Head with the ox he told them to Stand Back. He put the Ox in front of the Jam. And then Standing on the Bank Shot the Ox with a 303 Savage Rifle. The ox thought it was flies And began to Switch his Tail. And do you know That Ox Switching his tail forced that Stream to flow Backward And Eventually the Jam floated back also. He took the ox out of the Stream. And let the Stream and logs go their way."

Striding out to the West Coast after laying waste to the forests of the East and the Lake States, Bunyan dragged his peavey along the ground for a time; the result was the Grand Canyon. When the blue ox took sick out West, Paul and his loggers started to dig a grave for it. However, the ox recovered, and the unfinished grave became Puget Sound while the dirt heaped up next to it was the Cascade Range. In time, Paul built a sawmill to cut up an oversupply of Western logs. It was seven stories high, with a long bandsaw that traveled through the logs on all seven stories at once, and the twin smokestacks, made from Bunyan's shotgun barrels, had loggers stationed on top who brandished poles to fend off passing clouds.

And Paul certainly knew how to feed a man, too. The dining hall in his camp was so enormous it took waiters on roller skates 47 minutes to cross it, and in the kitchen there were gargantuan stew kettles in which the carcasses of cattle floated like chips in a millpond. To victual his loggers one day, Paul caught a flock of black ducks by cleverly laying his tarpaulin on the ground to resemble a lake, then quickly snatching up the four corners to bag the birds after they landed on it; the loggers ate the ducks, roasted crisp, with side orders of corn on the cob the size of a man's arm bearing syrupy kernels as big as buns, and finished the meal off with mountains of cream puffs as big as squashes. But Paul's crew was accustomed to such helpings. In his garden, it required stump-pullers to harvest the gigantic carrots and Paul was obliged to invent the steam shovel to dig up his potatoes.

The mightiest of loggers never died. When his coast-to-coast labors were finished, Paul Bunyan walked northward with his blue ox, to retire and take his ease in some green Alaskan valley. Behind him, however, the legends multiplied and were dressed with new embroideries, until the generations of woodsmen who

improvised them grew old themselves and died out.

The reality of the loggers' lives was filled with inglorious ingredients such as never encumbered Paul Bunyan. When they were laid off — which happened whenever a stand of timber was exhausted or the market for logs went into a slump — the loggers trudged from camp to camp looking for work. Or else they sought to buy jobs for a dollar or two at employment agencies, called slave shops, located in the sleazier parts of sawmill towns. Blackboards signaled what work was available: "Two fallers, $3.75 a day," "Donkey engineer, $4," "Bucker, $3." As a courtesy the "sharks" who ran the slave shops permitted loggers to store their bedrolls in a back room while shopping or carousing before lugging them off to camp. But, as one writer put it, "If by chance a roll was free of lice when it went on deposit, it was sure not to be similarly unencumbered when withdrawn."

As soon as they landed in camp, the cook told them where they would bunk and where to sit at the table; they meekly spread out their blankets and looked over the company they would be keeping. One observer, reporting on life in an early redwood camp, noted: "The striking thing about the California lumbermen is their diversity. Almost every European nationality is represented — French, German, Norwegian, Spanish, English, Scotch, and Irish, not to speak of Americans, Chinese and Indians." The Americans, the writer went on, "are often graduates of the Maine woods, or 'Bluenoses' from Lower Canada. These men are likely to become foremen, or sub-foremen, and form a nucleus around which the floating crowd is gathered. It often happens that a man will hire himself to labor in the redwoods who is fitted for a far better kind of work, but has met with misfortune."

Such a man might be the black sheep of a prominent Eastern family, a failed businessman, a disgraced professional man. A lumber camp was a good place to bury one's past, because few loggers had the bad manners to be inquisitive. Nicknames served as well as real names would have. There was Dirty Shirt Jim, known never to have changed his shirt or to have taken a bath; his camp mates forced him to do both one night, and by the next morning he had left in a huff. There was Celestial Sam, who carried around his dollar Ingersoll in a snuff box that had a hole cut out for the

A logger with a knack for tonsorial arts dispenses shaves on Sunday morning. Laundry was done *(right)* by pounding out dirt with a club.

stem; when asked the time of day, he would fish out the timepiece, announce the hour, and solemnly add: "And so much nearer eternity!"

One camp had a logger who was known to his mates as The Silent Man. "He was with us about two weeks and never spoke but once," one of his fellow workers recalled. The man would eat his meals in silence, work in silence, and in the evenings he would sit on the deacon seat by his bunk staring off into space. "He never wrote a letter, never exchanged a word with anyone. Finally, one night one of the men made bold to ask him whether he had been long on the Coast. Came the reply, 'My friend, I have been on the Coast a good many years, but I want you to understand that I am not going to undergo any cross-examination.' The next morning he quit."

Superstition thrived in the isolation of the brooding forests. Redwood lumberjacks working in Humboldt County, for example, were convinced that it was bad luck to start logging operations in a new area on a Friday; it was best to wait until Monday. They also believed that if a fat man came to work in the woods, three accidents were bound to happen in quick succession; better clear out. It was the woodsmen who may have originated the practice of wearing bands of copper around their wrists or ankles to ward off rheumatism. In addition, many a woodsman had his own private bugaboo. "There was this fellow who was with us three days or so," an old logger reminisced. "One evening as we passed the blacksmith shop he called our attention to the fact that the grindstone wobbled on its axis. We laughed. Next morning he rolled up his blankets and left, remarking, 'I never did have

no luck in a camp with a crooked grindstone.'"

Most of the loggers were bachelors—it was no life for a married man, at least not in the early days, when men liked to feel at liberty to float away whenever they grew tired of the boss, the work, or the food. Some camps experienced a 600 per cent labor turnover within a single year, and bosses were in the habit of claiming that they needed three crews—one coming, one working and one going.

Every logging camp knew a so-called short stake man, who never stayed long in one place. He was usually an eccentric whose unrealizable, pie-in-the-sky ambition was to find his paradise on earth: a logging "chance," or site of operations, where it never rained, where nobody ever got hurt and where the work, the pay, the food, the bunkhouse and the boss were all ideal. Because he was always on the move he was not unwelcome as a source of news. Logging historian James Stevens, who encountered the type many times, described Short Stake rather fondly as a man with a mania for personal bookkeeping:

"His prayerbook is the greasy, thumbed timebook in which is scribbled the figures of stakes (earnings) made on a hundred jobs, and his Sundays, evenings and holidays are spent in rapt porings over its pages." Hunched over his book, puffing on his pipe and clutching his chewed stub of pencil, Short Stake figures how he is doing on the present job. "He goes over his sums again and again to be sure that he has not erred in the company's favor. But he has it right the first time: eleven dollars and five cents he owes, and seven twenty he has coming in wages. Thus he is in bondage to the tune of $3.85. He sighs heavily; it is so hard to get ahead in this here world. But do not feel too sorry for Short Stake. He is the true bunkhouse optimist-dreamer. No camp has satisfied him yet, but he has no quarrel with 'conditions.' He has an ideal of a good outfit, and the dream leads him on."

Short Stake was worse off than his prayer book told him, and so were his fellow loggers in many camps. In good times and bad, a $5-a-week charge for bed and board was deducted from the workers' wages. The charge went on even when inclement weather halted work and pay, with the result that many men were chronically in debt to the boss, their accumulated earnings amounting to less than their bed-and-board bill. ◉

THE OLD MAN WHO LIVED IN A STUMP

Most loggers soon lost faith in the periodic attempts to exterminate lice from the camp bunkhouses. This bearded fellow, who worked for the Booth-Kelly Company in Wendling, Oregon, decided the only answer lay in keeping strictly his own company. Packing up his belongings, he moved into the hollowed out stump of a nearby tree. His plan was not a success; the bugs faithfully followed him to his one-man abode. But for a time he was something of a local celebrity—though his name is lost to history—and his home won special identity as the Wendling Stump, until it was blown up around 1900 to make way for railroad tracks.

Oscar Page's album: Memories of a domesticated logger

Oscar Page was a Missouri farm boy who headed West around 1898 to begin a logging career that would span almost 50 years. He shunned the rowdy back country logging camps for the more stable life of a married "homeguard"—or resident—logger, in Wendling, Oregon, a company town so dull that one worker grumbled: "You didn't have to know nothing to live there. They blowed a whistle for you to get up, they blowed a whistle for you to go to work and they blowed a whistle for you to quit."

Oscar thrived on the routine, mastering such jobs as faller, bucker, flume herder and river pig. But while these early photographs from his scrapbook, taken in 1900 and 1901, record a way of life that most footloose loggers scorned, they also evidence a calling where disaster was common. In a terse report of a friend's death, Oscar wrote: "A limb fell out of the tree that he was falling and knocked him on his axe and almost cut his arm off. There was no doctor and he lived about 18 hours."

On his 21st birthday in 1900, a duded-up Oscar strums his hybrid "harp-guitar."

Oscar *(third from right)* and friends pause during a Sunday hike. They went into the woods, Oscar wrote, "to see some large trees."

139

As the camp's first donkey gets up steam for a day of log harvesting, Oscar and a partner perch on springboards before sawing the back cut of a Douglas fir. One of his earlier jobs was with this donkey's crew.

A gallant Oscar serenades his bride Alice in front of their little cabin on the outskirts of Wendling, Oregon. He commuted to work in the nearby forests on foot, although some men erected cabins at the logging site.

An emergency crew of mill hands rushes to repair a leak in a drained splash dam while Oscar looks on from a high-and-dry boom log. Work continued around the clock as Oscar and his crew took over at night.

When the men did manage to get ahead of the game, they were generally paid in company scrip, or "orders," and when they got to town they often found that merchants discounted the scrip by as much as 20 per cent on the dollar. The townspeople were only being prudent; it was not uncommon for logging operators to go broke, rendering their scrip worthless. Or, out of sheer chicanery, operators sometimes refused to honor their own paper, leaving merchants holding the bag. Either way, the logger lost.

In 1886, a "gypo" logger (an independent contractor who supplied mills with logs) wrote a plaintive letter to the Port Blakely Mill Company about its failure to honor scrip.

"When men quit they must have something to show for their work," he wrote. The company had been quick enough to take away the logs his crew of lumberjacks had supplied, but as often as not it had ignored the chits that obligated it to pay the workmen. "Now I have logged for five years for you and if you cannot honor my orders for wages just say so and I will quit for good, right here. Have worked five years for nothing but a poor living and a pair of overalls once in a while that's all I have had. Inform me what you intend to do." There is no record of what, if anything, the company replied.

The only thing the men of the woods could count on with certainty was that they would engage in back-breaking labor from dawn to dusk and sleep like stunned sheep from dusk till the next dawn. These camps made few concessions to creature comforts; it was hardly that sort of world. Once a timber site was selected, log- or plank-walled buildings were hastily erected. The boss got a small shanty of his own, and the bulls, a dozen to 20 of them, got a shake-roofed, open-sided "hovel" to house them and their feed.

A typical main building in the 1870s was about 30 by 18 feet, with a partition in the middle separating the kitchen from the sleeping quarters for a score or so men. In the early camps, a circular fireplace made of rock and earth occupied the center of the sleeping barracks; above it, a crude wooden chimney was suspended from the ridgepole to conduct the smoke upward out of the room. (In later years, big iron stoves replaced the fireplaces.) Along the walls ranged the bunks, similar to those on shipboard, rising in two or three tiers. Some of the bunks were so confined that their tenants could only climb in from one end, thus the name "muzzle loaders." A thin layer of straw might cover the rough boards on which the men slept. In front of the bunks ran the long "deacon seat," a holdover, like the open fire, from bunkhouse construction back East. Made of planking or merely a split log, it was a bench to sit on and while away the off hours.

Such apartments were filled with an acrid mix of odors—sweaty, tired bodies, tobacco and snuff juice spat into the fireplace, rain-soaked work clothes hung on a line to dry. Suspended "tin pants," made of canvas and rubbed with pitch or grease to keep the rain out, were hung up on pegs or stood in corners. The pants were cut short for safety, so the bottoms could not catch on roots and trip up the wearer. Some loggers filled their empty high boots with water overnight, to keep them pliable (which may have been good for the leather but was terrible for the feet). Evenings, before the coal-oil lamps were blown out at 9 o'clock, the men would sit around telling stories or browsing through mail-order catalogs. A few of the freer spirits might dance together, stomping clumsily about; occasionally they would even shave—some men using the razor-sharp edge of an ax blade. But soon enough, most of the men—still wearing their long underwear and socks—were snoring in their bunks, dead to the world.

First man up at around 4 a.m. was the cook's assistant. Besides helping to get breakfast he had to revive the fires, refill the woodpiles and pour cold water into the men's wash basins, lined up on an outdoor shelf. A little after 5 o'clock the cook roused the sleepers—not gently. He could jangle them awake by banging on an iron triangle or, as in one 1870 account, "he walks to the door, puts a bullock's horn to his mouth, and blows repeated loud blasts."

An hour later, having washed and breakfasted, the men were heading out to their jobs. On many dark mornings they needed lanterns to find their way to work. They could only sense how cold it was outside because most bosses forbade thermometers in their camps. Not that it ever got so cold, on the mild western slopes of the Coast Ranges, that loggers would actually freeze when they stepped outdoors; but on a wet and clammy morning they could complain, if they had a mercury column to inspect, that 33° F. *felt* like 33

below. East of the Cascades it often did get down to zero and lower. The loggers remembered a Paul Bunyan tale about a winter so cold that the very flames of the cookhouse fire froze, and the men's oaths, uttered as they headed shivering for work, turned into ice and hung solidly in mid-air.

At least the loggers ate well in most camps. They had to, performing physical labor that could burn 8,000 calories a day, three times more than was used up by ordinary men doing ordinary jobs. The cook's long oilcloth-covered tables, with an inverted coffee cup set at each place, were piled high with food when the men filed in. The bill of fare was the same for breakfast, lunch and dinner, and a man could eat as much as he could hold of whatever he wanted. (In most camps, the men returned for lunch, but in some camps, when crews were working far out in the woods, the food was brought to them.) There was almost always roast meat, if only from an injured bull that had to be slaugh-

tered, along with hash and stew, a mess of greens, hot cakes and oatmeal, baked beans and potatoes, puddings and dried fruits, camp-baked bread, doughnuts, cookies, cakes and pies.

The cook, often a Chinese, in the California camps, was an important man with an endless and arduous job. A new cook was supposedly on trial for two days, but in reality his trial resumed at each and every mealtime. If a diner disliked the food, he could ask the boss to "make 'er out"—give him his pay so he could move on. An experienced logger who arrived in camp to start a job would be sure to size up the cook's pigs first thing. If they were too fat, that could be a bad sign—maybe the men were turning up their noses at the food. But if the pigs were lean and lively, that was all to the good—it probably meant they were not getting too many table scraps.

As long as he lasted, the cook's word was law on his side of the bunkhouse, which became a separate cookhouse as the camps became larger. By his fiat, mealtimes were silent, meant for stoking up and getting out, not for swapping stories and wasting time. The cook had his reasons: silence meant there were no complaints or quarrels within his hearing; he had little enough time to get the next meal ready; and besides, the boss wanted the men either at work or asleep, not dawdling around.

Over the decades the logging camps grew ever bigger and more complicated as the industry developed and became more mechanized, but loggers' appetites never diminished. A timber-company statistician at the turn of the century reckoned the daily food consumption of a 1,000-man crew in a major complex of camps, including the squads of cooks and helpers and such specialists as blacksmiths, saw filers and railroad men. They ate 1,000 pounds of fresh meat a day, 200 pounds of smoked meat, a ton of fresh fruits and vegetables, 900 pounds of flour, 600 pounds of sugar, 190 pounds of butter, 240 dozen eggs and enough coffee, tea and milk to wash it all down.

In the loggers' woods, a cacophony of sound assailed the ears: thwacking axes and rasping saws, clanking chains and roaring machinery—and above all, the rending crack and crash of trees shaking the earth as they fell. During working hours, only a tragedy was capable of stilling the industrious din. But that occurred with terrible regularity.

A Puget Sound lumberjack named Torger Birkelann remembered an afternoon when "everything seemed to stop. No whistles, no sound in the woods. After a while we saw men coming down, walking slowly and unsteadily. They were carrying something. We hesitatingly walked up to meet them and found it was Edvin, the whistle punk." The boy was only 16; he was part of a donkey crew, signaling the engineer by means of a long wire attached to a whistle on the donkey when it was time to haul in a log. Somehow he had been caught in the bight of the main line—a limp and harmless-looking length of steel cable that instantly snapped tighter than a bow-string every time the donkey pulled on it to drag in a log. He had been hurled into the brush.

"Tenderly carried in and laid on the nearest bunk, Edvin never regained consciousness; nothing could be done," Birkelann recollected. "There was no doctor within reach and it would not have done much good if there had been, for Ed's back was broken. Unashamed tears trickled down the cheeks of those rough-appearing, kindhearted men of the Northwest woods as we stood helplessly by and watched this 16-year-old boy pass into the Great Beyond. A sad day. No one felt like going back to work. 'Just a boy,' they said, while aimlessly walking around."

In 1896, a 25-year-old Swede named John Nordstrom was felling trees on a steep sidehill by Hood Canal. Another faller—also a Scandinavian—named Johnson was chopping away with him. Below them the sunlit waters, smoky blue as a Viking's eyes, stretched away to the north in a vista reminiscent of home. Sweating, the two fallers completed their cuts on a great fir. They shouted "Timber! Down the hill." The tree "talked" as its last fibers snapped, then leaned and toppled. Nordstrom leaped aside and then heard a cry.

"I turned around," he recalled, "and saw the tree carrying my friend with it down the hill. By the time I reached him he was nearly gone and I realized his back was broken. He lived only for a few minutes."

John Nordstrom helped carry Johnson's body back to the logging camp, helped fashion a rough coffin out of split cedar boards, and the next day he rowed 16 miles down the channel, the coffin balanced athwart the rowboat, to a point where a ship could take it aboard. He sadly accompanied the body to Seattle for the funeral, then left the woods for good.

There were hundreds of ways, none of them pleasant, to get hurt or to die. The loggers tried as best they could to look out for their own and one another's safety, just as miners, sailors and cowboys had to do at a time when no one else did. There were no unions or accident-insurance policies in existence, few safety regulations and inspections; and whenever a man or boy was injured or killed, his bosses had the tendency to blame it on his own inexperience or carelessness or bad luck. But even for the experienced, careful and lucky individuals, accidents lurked at every stage of the dangerous logging, milling and shipping process. When a mishap occurred, it generally went unreported in public prints—loggers were regarded as faceless, anonymous men anyway—or else it was recorded as a laconic squib in such trade journals as *West Coast Lumberman* or *The Timberman*.

Before his blade or saw ever bit into bark, the axman or bucker walking from camp to tree was in danger. Weighed down by heavy wedges, balancing an ax on one shoulder and a limber saw twice as long as his own height on the other, he could fall and be cruelly slashed if he so much as tripped over a root. While making his cuts in a huge tree trunk, he could lose his balance and tumble off his springboard eight or more feet to the ground. If the crown of his falling tree, hundreds of feet aloft, tangled with those of its neighbors, the trunk might not fall in a proper line; the severed base could leap backward, or twist and "walk" across its own stump, sometimes catching the faller before he could jump out of the way. An industry paper noted one of many examples: "Frank Blomquist was killed at work at camp. The tree, in falling, split and kicked back, striking the unfortunate man with such force as to cut his body in two."

If the crashing tree should happen to knock loose a branch high aboveground, the severed limb—often as thick around as a man's body—could plummet downward and explode like a mortar shell, its point of impact on the ground anywhere within a radius of a quarter of a mile. For some curious reason the loggers, although most of them were unmarried, called these large pieces of debris widow makers. They were deadly enough by any name.

"The carelessness of woodsmen is proverbial," another trade journal reminded the employers who read it. "A proof of this is at hand in Mason County, Washington. Two choppers were debating whether a limb that was swinging back and forth would fall. They realized that if it did, somebody would get hurt, however, and they decided to cut the tree down and trust to luck. They went to work, the limb fell and one man was instantly killed. They could just as well have cut down a score of other trees that were safe, but perhaps the element of danger decided the case."

A story not reported in any lumber paper, but passed along from one camp to another, told of a grisly suicide. A faller chopped out the undercut of a tree with his partner, then helped complete the upper or back cut. He said nothing about what he intended to do next. When the tree quivered and began to fall,

the partner leaped down to safety—and the suicide flung himself into the undercut, to be crushed as the tree came down.

Few jobs were as lonely or dangerous as that of the windfall bucker, whose task it was to saw apart wind-felled trees in order to get them out of the way of good timber purposely being felled. Nobody could tell in advance which way the logs might pitch or roll. And the bucker had to look out for more than the huge, multi-ton logs he was sawing through. Buckers preferred to work at least 1,000 feet away from a team of fallers, so as to be well clear when a tree crashed to earth. But that was not always possible, and the fallers were not always aware of how close to a bucker they were chopping. The bucker might hear a cry of "timber," hear the familiar swish of a falling tree and look up to see a giant Douglas fir coming straight for him at lightning speed. If the bucker failed to show up for the next meal, the rest of the crew might assume he had grown tired of his lonesome work and had cleared out. He might not be missed, or his body found, for days.

The high climbers, the spar men and pulley riggers on their lofty perches 200 feet to 300 feet above the forest floor risked a grim variety of accidents. A climber could sever his safety rope with a misdirected ax blow (to reduce that risk, ropes in later years were made with wire cores). A gust of wind could catch the falling top of the tree a man was topping and swing it around to brush him from his perch. Or the swaying trunk could split and spread apart, giving the high climber only a moment to drop far enough down the tree to avoid being pinched to death between his rope and the trunk.

Once in a while a high climber seemed to have a charmed life. Loggers on Washington's Olympic Peninsula never stopped talking about Haywire Tom Watson's incredible escape from death. One of his fellow workers, Tom Mansfield, recalled the circumstances: "I was working with him in the woods one day, and I saw him fall 120 feet, while going up a fir. He hit the mud feet first." The deep, soft mud cushioned Watson's fall and "when they drove him off, he was still smiling and waving his arms, as if in victory over death. The doc found nothing wrong with him, so he put his clothes back on and came to work." ◉

For a fortunate few, all the comforts of home

For most loggers during the pioneer period it was a primitive world without much room for the amenities of life. The lumber camps were often little more than bivouacs along the line of march through the forest fastnesses. But for those who worked near company headquarters or transportation centers, life was considerably more settled.

These "homeguard" operations were typified by the sawmill town built near Springfield, Oregon, in 1900 by the Booth-Kelly Company. Gone was the barracks-like bunkhouse of camps "out in the tules" — or backwoods. The unmarried men lived in a dormitory with 46 rooms, two men to a room; those with families got private cottages. The mess hall boasted cloths on the tables and waitresses to serve. There was a bakery and a general store, two reading rooms, a church and a schoolhouse—all lighted by electricity.

As time went on some companies even brought a few creature comforts to the men in the backwoods. By 1900, mattress ticking replaced cedar boughs in the bunkhouses of Sol Simpson around Puget Sound, and a crew of Chinese domestics kept the place clean. Women had begun to join the loggers in the deep forests as well. A lady named Florence Hills accompanied her husband on a log drive on the Willamette River in 1905 and kept a diary of her experiences. When one of her team of horses balked while fording the river, "Bill Eaton poured water in its ear," she wrote. "I guess we got out of there in a hurry. . . . Getting camp set up. Pete Tiolson killed a deer. I had a big venison feed. Fried venison, boiled potatoes, and gravy. Loggers will eat anything if it has enough gravy on it."

Perhaps her loggers would. But most loggers wouldn't; as a group they were notoriously persnickety about their food. "We have seen more trouble over a keg of poor butter," *The Humboldt Times* once observed, "than during the whole season over wages." And woe to the "boiler" (bad cook) — male or female. A favorite loggers' way of expressing displeasure with, say, a stack of leaden flapjacks was to nail the offending victuals to the cookhouse door.

Tables fixed, waitresses stand by to serve in a California camp around 1900.

With spare utensils hung in decorative array and flowers on the tables, an Idaho cookhouse staff proudly awaits a rush of ravenous loggers. The mustachioed cook wears his soiled apron as a badge of honor.

Lined up for a turn of the century picture with their dogs, kids in an Oregon mill town watch as the mutts react to a passerby's harmonica.

Female cooks operate a field kitchen for a small detachment of lumbermen in the Sierra wilderness near Mountain Home, California.

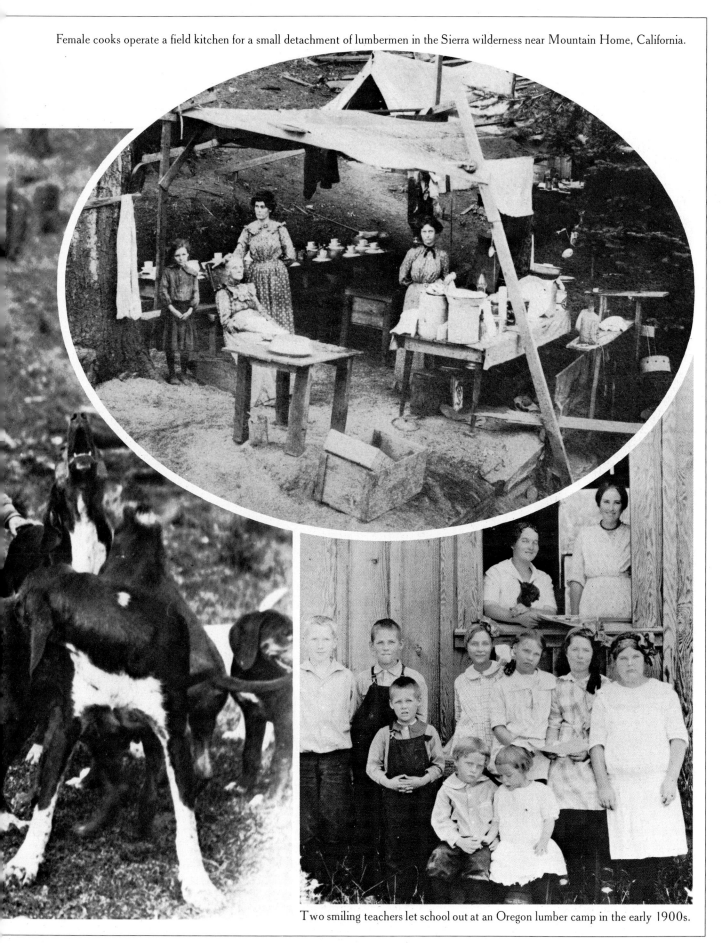

Two smiling teachers let school out at an Oregon lumber camp in the early 1900s.

But such miraculous reprieves were just that—miracles. "Consider the risks regularly taken by one group of workers, the chokermen," began a study of logging mishaps. The chokermen were responsible for fastening the steel cable of the donkey engine to a log so it could be hauled in from where the tree had fallen. "A chokerman and his partner, grabbing the steel cable while it was 'hot'—still swinging in the air—would crawl down through the limbs and chunks of trees on either side to complete hooking the choker under a log," noted the study. "At the same time, halfway in along the 'road' to the donkey, a sapling 100 feet high, weakened and undermined by all the logs slamming against it as they were dragged past it, would finally topple and fall across the whistle wire, breaking it and blowing one short, fatal whistle. Immediately the engineer would go ahead on the mainline, the chokers would tighten up, sinking into the bark; the log would leap out of the hole, crushing the bodies of the men beneath. Two men dead!"

Even after the logs went "down to splash," to wind up corraled in booms beside the sawmills, they were treacherous underfoot and could close over anyone who fell into the water between them. The calks in the soles of the loggers' boots gripped the rough bark of firs and hemlocks securely, but could slip on the stringy bark of cedars and spruces. Many men could not swim anyway and, once submerged, their heavy boots held them to the bottom like a diver's weighted shoes. They could try to walk the bottom toward shore—if they could remember, immersed in the dark water, which way the shore lay. A pathetic little item in a lumber journal told of one such accident:

"Two workmen employed in the Chehalis County Timber Company's camp drowned Saturday, April 28, in the pond. The bodies were discovered Monday, near the boom which forms the pond, tightly clasped in each other's arms. The unfortunate men had evidently tried to take a short cut to the camp by walking the boom sticks with the result that both were thrown into the water and drowned."

The donkey engine, the flume, the logging railroad, high-lead logging with spar trees—every advance in technology seemed to accelerate the accident rate. Donkey boilers blew up, people fell from flumes, runaway cars jumped the tracks spilling their logs like matchsticks. Finally, when a log rode up the conveyor from the mill pond into the sawmill, every piece of machinery that attacked it—from the big "head rig" saw at the entrance to the planers and edgers that finished off the boards—was a constant threat to the mill workers.

In the pineries of the Sierra Nevada, one out-of-the-ordinary hazard dated back to the Civil War. "The secession of South Carolina," lumber historian W. H. Hutchison reported, "cut off the vital 'naval stores' —turpentine and resins—that the Union needed to maintain its sea power in a day of wooden ships." California, a new and nonslave state, offered prizes to anyone who would help meet the shortage. "Soon hordes of men swarmed into the Sierra Nevada to tap the Ponderosa pine for pitch. When the Civil War ended, South Carolina quickly regained its former position and California's turpentine industry died out. The great holes that had tapped the bases of the pines were plugged with rocks to seal the wounds. Over the years, the rocks were gradually covered by healing growth"—and they played havoc with sawmill bandsaws that struck them decades later. Bandsaws were fearfully lethal when they broke. In one mill a broken bandsaw flailed around the floor until it curled itself like a snake into a corner, full of deadly tension and ready to strike out at any time. With great care, the mill hands poured cement over the steel band to encase and immobilize it.

Of all the sawmill workers, the men who rode the great head rigs, the log carriages that whisked timber back and forth against the saws, had the scariest jobs—and the most exhilarating. For hours on end, they worked within inches of a bandsaw that traveled at the rate of 10,000 feet a minute, and the machine they rode had the attributes of a mechanical bucking bronco. Historian James Stevens recalled his first job in a steam sawmill that cut pine in northern California, not far from Mount Shasta. Stevens, then 18, "was to have a tryout as a dogger on a head rig log carriage." The heavy carriage slid back and forth on rails, and carried the logs into a fearsome, endless bandsaw that was spun over 10-foot upper and lower flywheels and was cooled by streams of water.

"Naturally," Stevens wrote, "there must be some arrangement for holding the log firmly on top of the carriage, and some arrangement for moving the log over a few inches after each cut, so that there will be a new area for the saw to bite into. The 'holding down' device consisted of big, sharp claws — 'dogs' — operated through hand levers by men riding the carriage as it shot back and forth. That was my job. I climbed aboard. A signal came from the head sawyer to bear down. At the rear lever, I bore. Before my eyes the big claw bit down into the bark of a ponderosa pine log about four feet through. I locked my lever. In a split second, then, I was taking a death grip on that lever as the carriage seemed to plunge from under my feet." The next moment, as he rode ahead on the massive carriage, "the scream of the bandsaw hit my ears.

Shining steel and water flashed before my eyes, then a new-cut slab toppled and fell away from the saw. Then I was holding hard to keep the carriage from running out from under me as it shot back into position."

Clinging tightly to the carriage, he hurtled forward and backward the whole afternoon as the sawing went on. Before long, wet strings of hair were plastered over his eyes, and his legs and arms ached from the strain. And yet, "it all looked good to me. The uproar was music. The saws sang songs. As I got the swing of the log carriage, I seemed to sail along with the sawlogs. The opening through which the logs came into the mill was wide as a barn. It framed a picture made up of the water of the mill pond, which

A log boom on an Oregon slough is so securely locked together that the drivers have confidently let a woman, a baby and a dog visit them.

Relaxing as their campsite is dismantled before being relocated downstream, log drivers on Oregon's Willamette River wait, their peaveys ready, for the day's work to start. Known as river pigs, the men labored, they said, from "can to can't" — from the first light to see by until nightfall.

153

shone like silver, and the pine logs, which shone like gold. On beyond, green timber rolled away to the high mountains. Far away, the snow peaks were bright and sharp against the fall sky. The air was rousing and sweet to breathe. An hour before the quitting whistle I felt that I could eat a horse."

Like Stevens, most loggers felt that, with luck, they could cope with most hazards of the woods, and even take a sort of wild pleasure in their constant courtship of danger. But fire was the one peril that struck ter-

ror, pure and simple, into the heart of every woodsman. Once started, a forest fire was unstoppable except by an act of God, a catastrophe that could snuff out not one or two lives but many at a time. It was feared in the sawmills—few of which escaped at least one burning down—and with more reason in the woods, where incineration was harder to avoid.

The summer of 1902, which culminated in the worst fires ever known in the West up to that time, was hot and dry all over the Pacific coast region. High temperatures and low humidity had an ominous

Raging flames consume a shingle mill in Fall City, Washington. Fire was the loggers' worst enemy. Sawmills were particularly susceptible

cumulative effect on the forests, sucking billions of gallons of moisture out of the trees and the ground. Transpiration, the benign process by which trees abundantly exhale water and carbon dioxide into the atmosphere, was thus speeded up and thrown out of balance, leaving the forests bone dry and all too ready to burn.

On September 11, a searing, merciless wind from the deserts east of the Cascades and the Sierra swept over the mountains toward the Pacific. Everyone knew how the fires started: cinders from donkey engines and locomotives, sparks from stovepipes, spontaneous combustion. Dozens started at once, from Canada down to California. One hundred and ten separate conflagrations were counted; they consumed three quarters of a million acres of timber—an area as big as Yosemite National Park. Forests exploded like guncotton, the fires driven by the winds from behind and pulling in great upward drafts of air from ahead. Many fires "crowned"—speeding through the treetops like waves of breaking surf.

By one account, "men on the uplands could see them raging as far as the eye could reach: billowing

because of all the sawdust and scrap wood lying around ready to be ignited by a spark from clashing metal or a broken oil-burning lantern.

A first and last run over a not-so-perfect trestle

Early on the morning of November 25, 1912, a logging train chuffed out of the Seeley and Anderson camp in Coos County, Oregon, on the inaugural run of a new four-mile line linking the camp to the Coquille River. Pulling the three heavily laden log cars was the latest-model Shay locomotive. Aboard were three crewmen and four passengers from the camp—including two injured loggers going to a doctor.

En route, the rails spanned a canyon by means of a huge wooden trestle, almost 500 feet long and 100 feet high. The train started across—a little too fast, the engineer judged, and so he applied the brakes. But the momentum put a forward strain on the great structure. It started to creak and sway. The engineer only had time to cry out, "Boys, we're gone!" before the girders began to topple like dominos, one pushing over the next, until the trestle tumbled into the canyon.

Three of the men were killed outright. One, though badly injured, struggled free and went for help. The other three died shortly after being pulled from the tangled debris. Investigators laid the accident to the absence of lengthwise braces.

The wreck was one of the worst in the annals of logging railroads—a history heavy with gruesome disasters as the loggers recklessly pushed their hastily built lines through some of the most rugged country in America. There is no accurate count of the total number of casualties. But certainly many hundreds of people met their deaths in logging railroad accidents before 1912.

A bowed ribbon of track overlays a tangle of splintered beams from the collapsed trestle of the Seeley and Anderson railroad. At left is the edge of track at the point where it broke under the weight of the train.

Rescuers from nearby camps gingerly pick their way through girders surrounding the train's shattered engine. As it tumbled, the Shay locomotive spewed scalding steam from its boiler over the men inside the cab.

Grim-faced loggers, some of them carrying one of the victims on a stretcher, walk back along the tracks away from the wreck. It took 24 hours to free the last of the burned and mangled men trapped in the rubble.

rolls of smoke, fold over fold, white at the bottom, gray as they rose higher, and black at the top as they merged with the clouds. Seen nearer at hand, the arcs of fire were jagged scythes of flame eating into the wilderness. The leaping red-and-white tongues raced across roads, wiped out streams, encircled and drank up ponds. Roaring like tornadoes as they marched, the larger conflagrations scorched everything far to right and left of their paths. Bears, deer and wildcats fled frantically to clearings and lakes."

Loggers abandoned camp, farmers the fields they had cleared in the forests. In the smoke blanketed towns some were sure that this Dark Day, as it came to be known in the coastal forest belt, signaled the end of the world. To others it seemed the chain of Cascade volcanoes, from Mount Baker down to Mount Lassen, was erupting to doom them like the people of Pompeii and Herculaneum. Half an inch of ashes covered Vancouver and Seattle, which were dark as midnight at midday, and embers settled on Portland, 25 miles from the nearest forest fire.

A fellow logger told of Homer McGee, who was driving a team hauling a load of kindling to a cook-house near Mount Saint Helens, in southern Washington. McGee looked back and saw a fire. "Dumping his load he started the team back up the rutty road. The smoke grew thicker and the air hotter and suddenly at the bottom of a narrow ravine the full force of the racing fire came down with a deafening 'whush.' Even the stream seemed to be on fire. He tried to get the team turned around but had no time to do more than jerk the lines. He jumped headlong into the water already hot to the touch and buried himself as deeply as he could under a muddy overhang of roots. He prayed, living a horrible, searing lifetime in the minutes it took the worst of the blast to pass over him.

"The aftermath was something he always tried to forget but never could. When he dared crawl out over the hot ashes, he saw the charred horses and wagon iron. He groped his way along the edge of the river to the log landing, marked by the twisted piles of steel that had been the donkey engine. Some of the men had escaped. Eight others, including his two brothers, had been burned to shapeless heaps."

In the valley of the Lewis River, in the Cascade Range east of Mount Saint Helens, a party com-

The crew of a California logging camp in 1905 gathers to inspect the damage after their supply of blasting powder caught fire and exploded.

posed of 60 loggers, prospectors and farmers made their panicky way to a watery haven, Trout Lake. On logs and improvised rafts, they pushed off from shore into deep water, floating there for a duration of two days until the danger had passed. They were fortunate; in some lakes and ponds the floating logs ignited and burned up. Two families who were heading for Trout Lake did not make it. "A wall of flame came roaring with all the fury of a gale, trees exploding like gunfire and hurling burning brands in all directions. Death and annihilation came quickly. A set of iron wheel rims and a few black bones were all the evidence that life once moved here."

Not too far away, a mailman named Walter Newhouse was working in his yard when he noticed a fire spilling down a mountainside in his direction. "He lost no time in hitching his two ponies to the mail rig and racing down the road," an observer reported. "Yet he could not go fast enough. When a search party was able to get into the timber a week later, it found the horses in positions of flight but it was days before the body of Newhouse was found. He lay against a log in a gully, clothes and skin burned away, a short length of buggy whip in his hand." The wife of a neighboring logger, Mrs. John Polly, had "fought her way out of her burning house, baby in her arms, leading her young brother by the hand. The bodies were found in that relation."

Around the Saint Helens volcano area, the fire was remembered as the Yacolt Burn. As it happened, the railroad town of Yacolt was spared. Its people hurried off to a creek, kept themselves wet all night and returned to find their homes blistered but otherwise unscathed. The fire had simply leapfrogged over their hamlet, burning the surrounding timber but not pausing to consume Yacolt.

"From the Cascade Mountains to the Pacific the land was in darkness," remembered a historian by the name of Archie Binns, who had witnessed the fire as a child. "Down toward the Columbia River one forest fire raced through 250,000 acres of timber." Off the coast in the Pacific Ocean, "a lightship rolled at her mooring, with lights burning day and night and whistle blowing unceasingly. Ships with coughing and tearful crews groped their way toward the coast through a pall of smoke that extended forty miles to sea."

Towns in the path of the firestorm, said Binns, "disappeared forever and their names are forgotten. Families loaded what they could into their wagons and drove away, stunned and wandering aimlessly in the gloom of smoke and falling ashes. A mother and her children suffocated in a cave twenty yards from safety. And, as a final touch of terror, two desperadoes cruised in the pall of smoke, robbing and murdering and giving the slip to pursuers who were blindfolded by the dark."

Eventually, following seven days of inferno, the wind shifted around to the west, clouds rolled in from the Pacific Ocean and it began to rain. The fires were driven back upon themselves and then smothered, one after another. By September 17 the last of the conflagrations had died out.

The official estimate was 12 billion board feet of timber burned. Loggers went back into the charred forests and salvaged half a billion feet—with great difficulty, because dead timber hardens, resisting saws and axes as if it were petrified wood. The human toll was put at 35 deaths—astonishingly few considering the number of lives that were threatened. But no one ever really knew how many men, women and children died on the Dark Day, their ashes mingling with those of the cremated trees.

Bad record keeping, callous employers, undiscovered bodies all made it impossible to tally the casualties in the woods. It was said that a man's life expectancy as a logger was seven years—no more. By one educated guess, from 1870 to 1910 one logger died every other day, on the average. But that number—close to 7,500—was only a guess; the grand and horrible total, the cost in lives of extracting a treasure in timber for America and the world, would never be known.

One observer of the lumber industry, H. T. Hughes, a veteran logger himself who was fortunate to emerge from the game alive and well, at one point summed up Western logging as "a locus of crazy, wild industrial activity, a business carried on by shouting young men with nails in their boot soles, who traveled like smoke among crashing trees, dancing in a mechanical ballet whose accompaniment was the shrill music of shrieking steam whistles and the chugging of powerful engines. Civilization never before saw the like of West Coast steam logging. Only war compares to it." ◉

The different sound of Sunday

Sunday was the only day the camps did not ring with the cacophony of axes, saws and donkey engines. But the leisure hours were not always as quiet as is suggested by the newspaper readers below. From nearby woods there often came the crack and boom of gunfire as loggers turned hunters to eke out the cook's budget of fresh meat, or ran trap lines for valuable furs and game delicacies. There was, too, the sound of music, as woodsmen who had laid down their "Swede fiddles"—as crosscut saws were called—sawed away on real ones, or strummed and blew and thumped on other instruments while the rest of the crew vied for a chance to dance with the cookhouse ladies.

Oregon loggers display their favorite Sunday reading: the *Police Gazette*. Copies were passed around until the issues disintegrated.

Sharing the limelight with their hunting dogs, loggers from a Klamath County camp show off a bag of seven bearskins. Bear were taken in the fall, when the fur was prime. Steaks and ribs were roasted; hams and shoulders were smoked.

John Sells (violin at rest, second from right), a donkey engineer, rehearses some tunes with his lumberjack band at Simpson's camp on the Olympic Peninsula. Apparently the band was completely unaware of the nonfiddler on the roof.

165

Three logger-trappers stand surrounded by the trophies of their Sunday expeditions and by Klamath Indian artifacts taken in trade for furs.

5 | Blowin' 'er in on Skid Road

A logger's play was as strenuous as his work—and sometimes just as dangerous. His spare-time pursuits were elemental: roughhousing, drinking, gambling, womanizing—and the closer to hand, the better. To provide him with his pleasures, small towns sprang up like toadstools. No matter that the town—like Grand Forks, Idaho, at right—might be a few shacks in a clearing, or even a huddle of tents; all he cared about was falling into the first drinking and wenching spot he could find.

Booze was the breath of life to him, and so urgent was his thirst that he would down anything from grain alcohol to horse liniment if nothing better could be had. Boozing led to brawling of a monumentally ferocious kind. In the rage of battle, men ripped away opponents' ears and chewed hunks from their hides. "The Marquis of Queensberry would have been shocked," reported an observer. "When a man was down, his opponent jumped upon him with both feet, kicking and tearing at him with the cruel calks in his shoes."

Somewhat more varied were the recreations awaiting the loggers in the booming lumber cities. To edify the men from the woods, one Seattle saloon, called The Palace, kept on display a big glass jar full of writhing snakes; and the Humboldt in Aberdeen was crowded with fascinating curiosities ranging from Indian tribal garments to mastodon bones. For the logger in search of gentle companionship, Aberdeen also featured a street of high-class bawdyhouses bearing such names as Harvard, Columbia and Yale; the loggers thought it very educational and referred to it as College Row.

In big towns and small, a man's money disappeared as fast as grub on the cookhouse table. "Many a fool lumberjack," said one, "spent his whole winter's work by noon the first day in town. Ho hum! Some palmy days!"

Crammed into a clearing deep in the timber of northern Idaho, the lazy-looking little town of Grand

Forks — its main street a simple swath through the forest — was in fact a notorious collection of bars and bawdyhouses patronized by loggers.

Loggers lounge in a well stocked, not very tidy Mendocino County saloon in 1900. True to type, the bar featured a brass rail to withstand the

loggers' calked boots, spitoons (often ignored), portraits of feminine pulchritude and masculine muscle, and an implacable "No Credit" sign.

Demurely proffering their charms in an establishment called The Club this bevy of "soiled doves" welcomed loggers and their cash to Tacoma,

Washington, in the 1890s. Sometimes clients paid in scrip, obliging Madam and her girls to spend their remuneration at the company store.

Time out for booze, bawds and brawls

The publishers of *Godey's Lady's Book*, a Philadelphia fashion journal of the 1880s, were delighted if a little surprised. Subscriptions were pouring in from Vancouver Island, a place so unfamiliar that some of the staff had to look it up on the map. There it lay, a continent away, far out west and to the north of Washington Territory's Olympic Peninsula. The location touched off smiles of satisfaction; if the fame of their magazine had spread all the way to Vancouver, clearly they were doing something right.

On Vancouver, the new subscribers were equally happy. Indeed, they were sitting around the bunkhouse slapping their thighs and snorting with glee. To the primeval forest where they labored and lived, fortune had directed a rare stroke of luck. A bevy of determined camp followers had set up shop within walking distance. These ingenious damsels had trekked into the wilderness prepared to flourish against all obstacles. Armed with the knowledge that company rules forbade the presence of loose women within camp limits and that advertising their presence would invite banishment, they had outfitted themselves as magazine saleswomen peddling a number of publications, including *Godey's Lady's Book*. New subscribers, they explained to the eager loggers, could pick up their receipts at the ladies' cabin just down the road where a warm welcome awaited them. At that moment the lumberjacks of Vancouver Island made a commitment to periodical literature that became legend among loggers.

It was just the sort of yarn to make a logger heave with laughter when he heard it—and hear it he cer-

tainly would, for it was guaranteed to enliven a bunkhouse on a cold, wet night. Other than tall tales, there was precious little in camp to create diversion. Hunting or fishing, maybe, on Sunday. But women? Never —or as good as never. Whiskey, which the logger yearned for almost as much as he did women, was also hard to come by, though in violation of the rules loggers did manage to smuggle in bottles, and occasionally a shanty built on the fringes of the camp sold them rotgut. Brawls there were in plenty, but somehow they did not palliate a man's primal hungers.

He needed something more: bright lights and hurlyburly, the noise and excitement and milling confusion of "town." On a clear day, a logger perched atop a Douglas fir in the Cascade Range could actually see Seattle, tantalizingly close as the crow flies. On foot, however, he might need all of Sunday just to get there. And so he lay in his bunk and bragged and dreamed: of the girls he would crush in his bear-hug embrace, the booze he would down, the skulls he would crack, the general hell he would raise when he finally hit Skid Road to "blow 'er in!"

Skid Road, capitalized, had come to mean those brightly lit blocks in any logging town that were lined with saloons and honky-tonks, cheap restaurants and lodging houses. In later times and other places the term was corrupted to Skid Row, but not by anyone who knew the derivation.

On Skid Road a man could cut loose and exorcise all his pent-up feelings of boredom and loneliness. The first to bear the name was Mill Street in Seattle, later called Yesler Way. Originally Mill Street had indeed been the skidroad down which logs slid to Henry Yesler's sawmill. The early logger bent on pleasure simply took the same path downhill on which the oxen and logs had preceded him. At the bottom—where the town that had grown up around the sawmill was waiting to

---

Loggers on the town in Gardiner, Oregon, memorialize the occasion before "blowing 'er in" at the local pleasure spots. After three or four months of work a logger might save more than $100 for a grand spree.

separate him from his money — he might be in for some hazing about his "skidroad gait": the peculiar shuffle-and-hop loggers developed while matching their stride to the spaced cross-logs on the long way down.

The Western loggers' preferred hangouts — Yesler Way in Seattle, Burnside Street in Portland, Hume Street in Aberdeen, Idaho's roaring St. Maries and St. Joe riverfronts — all dangled the same gaudy lures as had their counterparts back East: Haymarket Square in Bangor, Maine, or Sawdust Flats in Muskegon, Michigan. But as if to match the vaster timber of the West, what the Skid Roads offered was bigger, wilder and more spectacular.

Knowing the enticements that lay ahead, any woodsman with an ounce of prudence in him made an effort to tend first to his humdrum needs, before his money ran low and his good resolutions ran out. If his teeth were bothering him, he headed for the local office of Dr. Painless Parker, a dentist who ran a thriving chain of tooth-pulling parlors. Painless Parker's man, as often as not, stood in an open doorway demonstrating his professional skill on a shill, dressed as a logger, whose calm demeanor and obvious good spirits were designed to disarm the timorous.

After his teeth were fixed, the 'jack might get a 25-cent haircut, preferably from a lady barber, or if he wanted to save money he visited a barber college where he might be shorn free of charge. Then he stopped at the outfitting store to replace his boots or to buy a cheap Sunday suit, often of the vibrant purplish blue favored by his peers. In this resplendent state he was ripe to sit for a photograph to be sent back home.

Along Skid Road a freewheeling logger might next be tempted by the tattoo parlor, or by the phrenologist who divined character and fortune by the bumps on a customer's skull. Shooting galleries where a marksman could show off his skill abounded, and those who fancied themselves as poker players could find games galore — along with schools of real card sharks to match wits against. Sprinkled among these enterprises were more decorous havens of rest and recreation, such as the missions that sold salvation and gave away soup.

But none of these diversions really qualified as "blowin' 'er in!" For that, Skid Road provided a sumptuous array of pleasure palaces designed to pop the logger's eyes, make him feel as powerful as Paul Bunyan,

This cluster of cabins at Martin Creek in Washington's Cascades offered beds and booze to loggers and others who could not get to Portland or Seattle. The woods were full of such "entertainment centers," but they lasted only until the area's timber was cut and the customers moved on.

Seattle, a rough-and-ready logging town, exudes a deceptively pastoral air in this 1874 watercolor. The street running from left foreground to the waterfront featured a red-light district called "Down on the Sawdust" because it was originally located on land fill made from sawmill wastes.

relieve him of his tensions and his cash, and bring him back for more. Lumberjacks all over the West agreed that of the thousands of saloons in their ken, one stood out as absolutely the grandest of all. That was Erickson's, on Burnside Street in Portland's rough, tough North End. A sign over one door proclaimed it to be "The Workingmen's Club." It was plebeian enough on the outside—an assemblage of dingy frame structures occupying most of a city block. But the moment a logger shouldered his way through any one of its five swinging doors, there before his dazzled and delighted eyes lay nirvana.

Running in a huge rectangle around an immense room was a bar that had to be the biggest monument to alcohol in the entire world—684 linear feet of mirror-polished mahogany, and every foot of it jammed with happily imbibing loggers. ("Bigger than Erick-

son's bar" became a woodsman's standard term for anything so huge it beggared description.) On one wall over the bar hung a tapestry-sized oil painting that a man could gaze at rapturously for drinks on end. Entitled *The Slave Market,* it was an opulent work of art populated with voluptuous pink nudes who were captured by the legions of Rome. The main room also boasted a concert stage, and while taking in the singing and high-kick dancing of Erickson's chorus line, a logger could tap a boot to the rhythm of an eight-piece ladies' orchestra. Between acts the building vibrated to the chords of a "$5,000 Grand Pipe Organ."

There were dozens of gaming tables where a customer could squander his stake. Around the mezzanine level ran a row of cozy, curtained booths where he could do his drinking, dining and love-making in private. The girls who kept him company on the mez-

zanine were freelances, not on the house payroll, and it was a strict rule that none of the girls was to set foot on Erickson's sacrosanct main floor.

One of the things loggers liked best about Erickson's was the gargantuan free lunch that was available day and night. It was strategically deployed along the bar, and as long as a patron kept buying drinks he could help himself. August Erickson, the blond blue-eyed Finn who founded the saloon in the early 1880s, had been lavish and tasteful in setting up his lunch counters. They bulged with quarters of juicy roast ox, thick logs of sliced sausage, stacks of bread cut a generous inch-and-a-half thick, blocks of aromatic Scandinavian cheeses, vats of herring in brine, quart jars of burning-hot, homemade mustard.

"A dainty lunch," Gus Erickson called this repast, knowing that the fame of his inexhaustible food, the quality of his liquor and the glamor of the surroundings would draw enough trade from the woods and the seas to make it all pay.

The whiskey was poured, at two shots for a quarter, from bottles displaying the proprietor's handsome likeness, complete with dark suit and pince-nez. Among Erickson's appreciative customers was a logging-camp blacksmith from Denmark named Charles Oluf Olsen, who later turned writer and fondly described "the delicious concoctions Erickson's accomplished bartenders could conjure from the mysterious-looking bottles on the back bar. On frosty mornings there was a certain chill-chaser at the making of which one of the drink-dispensers was a wizard; a thin glass, delicate almost, half full of boiling water, a silver teaspoon of powdered sugar, a generous dollop of Jamaica rum, a touch of lemon peel and the merest dash of

nutmeg." The result was a pick-me-up, wrote Olsen, "that would have thawed out Paul Bunyan in the memorable Winter of the Blue Snow!"

Erickson's squads of bartenders and bouncers were as renowned as his whiskey. The bartenders were imposing, mustached men, uniformed in spotless aprons and white shirts with elastic armbands. Polite and attentive, they were ready to discuss anything except religion and politics. Neither they nor the bouncers objected to mere boisterousness; the Workingmen's Club was a success precisely because it was the kind of place where loggers and seamen could bellow at one another to their hearts' content. But if anyone was of a mind to fight, he had best step outside. Let a guest start a brawl, or even threaten to, and the sharp-eyed bouncers moved in. If a warning word failed to cool off the hot-tempered customer, whatever his size might be, he was manhandled through the nearest pair of swinging doors in a matter of seconds to land in a heap on the sidewalk.

The biggest and most efficient of Erickson's bouncers was a 300-pound behemoth by the name of Jumbo Reilly. Jumbo did not even bother to trade punches with a drunk-and-disorderly logger; he simply leaned heavily on the miscreant and, marching majestically forward, swept him unceremoniously out through the doors. A favorite story in the lumber camps was told of the time Jumbo ejected a cantankerous Swede known as Halfpint Halverson. Landing outside Erickson's Second Street entrance, the well-soused Halverson picked himself up and wandered around the corner to find another saloon. What he found was Door Two of Erickson's, on Burnside Street, and once more Jumbo tossed him out. Gradually making his woozy way into — and quickly out of — Entrances Three and Four, Halverson finally fetched up at the fifth door, on Third Street. Jumbo was there waiting for him. "Yesus!" groaned Halfpint, "iss yu bouncer in every place dis town?"

No wonder lumberjacks loved Erickson's saloon: everything about it was big enough. It became a cross-roads of the Northwest, and it was said that if a woodsman spent enough time there, he would hear tell of — if not actually encounter — everyone he had ever known. At the least he could always leave word that he had passed through; as an extra service, hundreds of messages to long-lost friends were tacked up on the club's commodious bulletin boards.

Erickson's was only two low-lying blocks from Portland's Willamette River, ordinarily a placid enough watercourse. But early in 1894 the Willamette, swollen with rainwater from the Cascade mountains, went on a rampage. Along with much of the city's North End, Erickson's was flooded and isolated. The resourceful proprietor hired a large houseboat and anchored it outside his saloon, in the middle of Burnside Street. Stocked with kegs of beer, cases of whiskey and the free lunch, the floating saloon did a lively business for several days until the flood waters receded. Patrons came by rowboat and raft, and some loggers paddled their way to the houseboat astride stray logs. For safety's sake, if not for sobriety's, many of them stayed aboard until the ark grounded.

The only other loggers' hangout that deserved to be mentioned in the same breath as Erickson's was Big Fred Hewett's Humboldt Saloon, in the Grays Harbor town of Aberdeen, Washington. By the early 1900s, Aberdeen and its twin, Hoquiam, were among the world's busiest lumber-shipping ports — and a highly dangerous pair of towns for the loggers and seamen who came there. The Humboldt, which Big Fred operated on hard-boiled but fair-and-square lines, grew famous as the one refuge among the area's 50 drinking places. "A logger's life was perfectly safe inside its doors," said one timberman of the Humboldt. "No gambler could bilk him of his hard-made earnings. No painted dancer could spill chloral in his whiskey glass. No plug-ugly could zap him behind the ear and then frisk his pockets. If after a few drinks the logger himself began to yearn for battle, he soon found himself bouncing over the sidewalk outside, with Big Fred's voice roaring behind him: 'Come back when you want

A 1905 advertisement for Erickson's neglects its most noteworthy feature: the immense bar that measured more than 200 yards and was staffed by 50 bartenders.

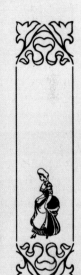

Erickson's Cafe and Concert Hall

THIRD STREET ENTRANCE

LADIES' ORCHESTRA

Entrances

26 N. THIRD   21-23-25 N. SECOND
245-247 ½ BURNSIDE ST.

Equipped at a cost of $130,000.00

AUGUST ERICKSON
PROPRIETOR

AUG. ERICKSON

181

to be decent! I do all the fightin' for this place!'"

A six-foot-one-inch hulk of a man, Hewett was himself originally from Maine, and so it was perfectly natural for him to indulge the lumberjacks. At Christmas time and the Fourth of July, when thousands of woodsweary loggers descended on Grays Harbor to mingle, drink and fight with mill hands, longshoremen and sailors from lumber ships, Big Fred stayed open around the clock. He cashed the timber beasts' paychecks and company scrip at face value, and stashed away their bankrolls for safe-keeping; his safe often held as much as $20,000 of their hard-earned cash. When a man who had entrusted his hoard got drunk and changed his mind, Hewett was sternly unsympathetic: the logger would get no money — not even to spend over the bar at the Humboldt — until he had sobered up.

Anybody could decorate a bar with billowy nudes; Big Fred's taste ran to more intellectual displays. Arrayed along the Humboldt's walls were cases crowded with coins, rocks and minerals from many lands, stuffed reptiles and birds, firearms, scrimshaw, Indian artifacts, portraits of Jesse James and Lillian Russell, even a meerschaum pipe on which the outlaw Cole Younger, one of the James gang, was said to have puffed away his days in a Minnesota prison. "My museum," was the way Fred proudly described his place, and his collection of curiosities was valued at $20,000.

Two loggers once reported an unusual find to Hewett. Beside a stream 500 feet from the nearest road they found a very old and crude Indian dugout canoe, covered with moss and lying beside the stump from which it had been hewed. Big Fred was interested; he offered the men five gallons of whiskey to haul their discovery, along with the stump, to his bar. When the massive memento reached his saloon, he sent off a piece of the stump, which showed the scars of a chopping tool, to the Smithsonian Institution in Washington, D.C. For years thereafter the canoe and stump were a prized exhibit at the Humboldt. A letter from the Smithsonian attested that the canoe was hacked out with native implements, probably before 1792 when white men brought the first steel axes West. Fred liked to say to customers, "Shows what a fellow could do with a stone axe when he got up a sweat."

Fred was never inhospitable, but the rare "dude" — a fancily dressed fellow from out of town with no cal-

No logger ever lacked for entertainment at the corner of Seattle's Occidental and Washington streets. After drinking his fill of good Bohemian beer—or whiskey—at the Our House Saloon, a man could wind his way upstairs to the 600-seat Lyric Theatre and see a rip-snorting burlesque.

luses on his hands — who wandered into the Humboldt was not made to feel entirely at home. On one occasion a dude bellied up to the bar and asked Fred to fix him a Manhattan cocktail. Hewett eyed him, then placed a big beer mug on the bar. Into it he poured a slug of his double stamp whiskey, a shot of gin, another of rum, a dash of brandy, one of bitters, one of aquavit and enough beer to fill the mug. He slid the concoction in front of the customer, gave it a leisurely stirring with his forefinger and said, "There, mister, is your Manhattan." When the stranger protested that this mess wasn't what he'd ordered, Fred rumbled ominously, "Drink 'er down." The man obediently did a bottoms-up, plunked down 50 cents and hurried out never to return. Loggers at the bar drinking straight double stamp thought that Fred had handled the intruder just right.

Though Fred never had to shell out money for wholesale rebuilding, as often happened to other saloon owners who could not control their rambunctious patrons, the Humboldt did suffer predictable wear and tear. Twice a year, after the midwinter and midsummer carousing, the wood floors had to be replaced: the spikes from thousands of calked boots turned the flooring into a mass of splinters.

But there was no way of talking loggers into leaving their boots in camp. A logger without his "corks" felt as uneasy as a cowboy without his ten-gallon hat. "Barefoot," in woods jargon, did not mean shoeless, it meant calkless. A smart saloonkeeper like Big Fred tolerated the boots as long as the lumberman's money offset the damage done by his spikes. Many saloons did, however, find it necessary to install rails made of brass in front of the bar. A man wanted to drink standing up, with one foot planted firmly on a rail. No wooden rail could last long under the impact.

Loggers insisted on wearing their boots to town not merely because they felt comfortable in them; they reasoned that they might need their sharp calks for protection. After a well-behaved beginning at, say, the Humboldt, a lumberjack with his steam up might go on to the notorious Palm Dance Hall, which Skid Road connoisseurs rated as the Northwest's wildest playground, famous for its 40 dancing partners and the raging, all-out fights that passion for them provoked.

It was in these fights, some of them grim indeed, that the calks came into play. When loggers fought, they eschewed knives or pistols; their weapons were bludgeon-like fists for smashing, thumbs for gouging and calked boots for stomping the will out of an opponent. A man who came out second-best in such a fight bore the visible marks of defeat for a long time. Loggers said he had a case of "lumberjack smallpox." In a serious brawl a combatant would also throw anything not tied down — chairs, tables, bottles, or other loggers. During one fine free-for-all, at a moment when a muscular axman stood poised with a helpless adversary held over his head like a railroad tie, an onlooker shouted a memorable piece of advice: "Don't waste him; kill the fiddler with him!" Many a saloon piano player learned to shelter himself from the fray (or his critics) behind a wire barrier; if anyone wanted a request number, he had to toss his money over the mesh.

A logger shrugged off the physical hazards of these alcoholic donnybrooks; the dangers weren't any greater than those he faced daily in the woods. But some Skid Road bars in the big lumber ports held a further, far more evil menace: the "crimp."

The crimp was a recruiter-by-force, a man who shanghaied other men to fill out the crews of departing ships. Hated though he was, he filled a certain niche in the economy of timber. Lumber camps by the thousands and sawmills by the score required ships by the hundreds to carry their wood to the ends of the earth. At every port from San Francisco to Victoria, sailors of all nationalities jumped ship. They had good reason. Compared to shipboard life at $20 a month "and slops," life ashore on the West Coast, at any job or no job, was infinitely attractive. And in the wide-open towns there was little likelihood of being rounded up and deported. Skippers whose vessels were loaded with boards and ready to sail thought it was well worth $35 to $50 a head to fill out their crews with able-bodied men. Their ships lost money lying idle in port. The crimp was a useful middleman, prowling the Skid Roads in search of victims he could tranquilize, abduct and "sell" to impatient ship captains.

Many Skid Road joints were made to order for the crimp. Like the waterfront sawmills alongside which they had grown up, the sleazier saloons and flophouses were built over the tide flats. In their back rooms were trap doors that opened over the dark, lapping waters of rivers or harbors. In such infamous places as the Bucket of Blood saloon in Everett, Washington, which was constructed on stilts over the Snohomish River, trap doors dropped away many a man who had been knocked out by chloral hydrate drops or anesthetized by too many drinks. If he had been marked by a crimp, he was fished out and delivered C.O.D. to a lumber ship. If no ship needed him, he was robbed of his stake and left down there—to drown or, if the tide was out, to crawl ashore when he came to.

As a rule the local police took no great interest in crimping. There was little pressure to stop the practice —and considerable pressure to wink at it. The trade's practitioners often operated behind a legal façade. Limey Dirk, for instance, was widely reputed to have grown rich in the kidnapping business while operating the Sailors' Boardinghouse in Port Townsend, at the Puget Sound end of the Strait of Juan de Fuca. The authorities never did anything about it.

In Portland the unsavory traffic was especially well organized: crooked bartenders and waiters, pimps and their women joined forces with the crimps and got a split in the fees. The flashiest crimp in Portland was Jim Turk, another boardinghouse proprietor, who affected a silk hat and swung a gold-headed cane as he made his rounds of the North End.

Turk's most notorious rival in the business was Joseph "Bunco" Kelly, pint-sized, unobtrusive, shrewd as a mongoose and a figure of legend. In the North End saloons it was said, and loggers believed it, that one dark night Bunco Kelly, hard up for live bodies to fill an order, wrapped a cigar-store Indian in a shroud and ferried it to a ship about to head out to sea. Kelly assured the captain that once the big lumberjack inside the tarpaulin sobered up, he'd make a sturdy and complaisant deckhand. "Nothing like some good sea air for a logger," Bunco liked to say. Eventually, however, the trapper was himself trapped. Bunco was caught redhanded by one of Portland's few incorruptible cops and wound up in the Oregon penitentiary.

Perhaps the master crimp of them all was Billy Gohl of Aberdeen. Thanks largely to him, the rainy Grays Harbor city was known as a port of missing men. Billy had drifted into Grays Harbor after going broke seeking his fortune in the Yukon gold fields. He made a certain name for himself touring the Aberdeen saloons with a grisly yarn about how once, while starving in Alaska, he had been forced to murder a man and eat the corpse. Billy looked tough, talked tough and acted tough. Starting out as a Grays Harbor barkeep he soon graduated to a more prestigious and powerful job as agent for a local sailor's union, with an office over the Grand Saloon on Heron Street. Not only was he in the body-snatching business, but he and his attendant bullies also forced their way aboard ships to shake down the captains in the name of the union.

One captain who especially loathed Gohl was Michael McCarron, master of the lumber schooner *Sophia Christenson*. One day, as the vessel, with a cargo of lumber, glided down the Wishkah River, McCarron caught sight of the bright red outhouse that hung over the river behind Gohl's second-floor office. On impulse, but with perfect timing, he swung the wheel. The long jib boom of *Sophia* swung out and knocked the privy off its supports into the water. McCarron's satisfaction was not quite complete; Gohl was not occupying the outhouse at the time.

Despite Gohl's record as a murderer, crimp and extortionist, he continued to operate and thrive, until his own carelessness brought his downfall. It was his usual practice, after killing a man for his money, to take the body by small boat out beyond the Grays Harbor bar and dump it where the currents would carry it off to an untraceable burial at sea. One winter this course seemed to be too much trouble and so Gohl took to depositing his corpses in the Wishkah and Chehalis rivers, where they were found floating under piers and in the log booms. Forty-four bodies were recovered, and enough of them were traced to Gohl to send him to prison in 1910. Soon after, workmen installing sewers and water mains on the Aberdeen waterfront began uncovering skeletons, and these also were attributed to Gohl's activities. Loggers and sawmill workers figured that this worst of crimps must have run up a death tally of well over 100 men.

While the Northwest's Skid Roads never lacked liquor and excitement and danger, until the 1860s they were conspicuously short of women. Many settlements near the sawmills had a ratio of only one woman to every ten men, and that one in ten was usually a pioneer wife who had crossed the Plains with her husband and family. In early-day Seattle, two men set out to redress the sexual imbalance. One was an amoral entrepreneur, the other an upright idealist. As was so often the case on the frontier, the entrepreneur had the only truly practical idea.

John Pennell, who had learned all the tricks of pleasure-for-profit on San Francisco's Barbary Coast, set foot on Mill Street in Seattle in 1861. He had come up from California on a lumber schooner, and one look around confirmed his predaceous suspicion that there was easy money to be made from the loggers and mill hands of Puget Sound. Within a month, Pennell built the first Skid Road bawdyhouse, the Illahee (Indian for home place). Only a few minutes' walk from Yesler's sawmill, it was simple and functional: an unpainted frame building with a dance floor, a bar and small rooms opening off one side. Pennell's musicians—violinist, accordionist and drummer—all came up from San Francisco, but the rest of his staff was recruited locally. It consisted of Indian girls, procured from their willing fathers in return for Hudson's Bay blankets.

In the Salish Indian culture of the Northwest the girls not only went unwashed, they groomed their hair with fish oil and, sometimes, urine. Even a brawny logger—in all likelihood, none too sweet-smelling himself —recoiled from these women except as a last resort. But Pennell saw to it that his Indian girls were scrubbed, perfumed and dressed in gay calicos. They could be danced with, after an ungainly fashion, and a lumberjack was expected to buy himself and his partner a drink after each dance. Hers looked like whiskey and was priced like whiskey but was really cold tea. A visit to one of the little rooms with a dancing partner cost the patron from $2 to $5.

The Illahee prospered mightily and in 1864 Pennell augmented his stable, importing a flock of "Frisco lilies"—girls from his old haunts on the Barbary Coast. His fellow citizens took to calling his establishment either the Mad House, for the wild goings on, or the Sawdust Pile, for its proximity to the mill (a girl who worked there was said to be "down on the sawdust"). But they had to admit that it brought business from all over the Northwest. And with the quick success of the Illahee, sin in Seattle became a major industry.

Asa Mercer, the high-minded 22-year-old president (and sole faculty member) of the infant University of Washington, had also noticed that Seattle was crowding up with womanless men. But Mercer had quite different ideas about how to remedy the situation. Back East, the casualties of the Civil War were creating a surplus of eligible womanhood. Why not, he reasoned, solve both problems at once? Mercer decided to go fetch brides for the men of his sawmill town.

The first of his two trips East went well. He signed up 11 Massachusetts maidens in the textile town of Lowell, which was depressed for want of Southern cotton. The ladies traveled west across the Isthmus of Panama, took passage aboard a steamer and received an ecstatic welcome to Seattle at midnight of May 16, 1864. Mercer's seamstresses and teachers, the fair and the homely alike, were whisked to the altar by eager grooms, and on the strength of his coup, Mercer was elected to the Territorial Legislature.

Soon he was planning a second, more ambitious trip, this one to bring 500 Eastern maidens to the wilderness. He talked a number of men in Seattle (only he knew how many) into paying him $300 fees for sight-unseen wives. He also planned to enlist the support of a powerful ally: as a child in Princeton, Illinois, he had been dandled on the lap of Abraham Lincoln, county lawyer and family friend. With Lincoln's help, he explained now, he hoped to transport the 500 women free on a United States Navy vessel; after all, his enterprise was clearly in the nation's interest.

This time everything went wrong. Shortly before Mercer was to see the President in Washington, D.C., Lincoln was assassinated. The Quartermaster General was unsympathetic and ruled that such use of government property as Mercer proposed was illegal. James Gordon Bennett of the *New York Herald* scared off hundreds of prospects by branding the mass-marriage broker a white slaver. Nonetheless, after nearly a year's delay Mercer got back to Seattle in May 1866 with 46 women, most of them young.

He had no way of parceling out the girls to all the men who had advanced him money; the latter certainly numbered well over 46, whatever their exact total. Nor did he have any way of refunding the advances: his year back East had cost him every dime of the money. But he somehow managed to pacify the investors and, on the night of May 23, he actually had the courage to get up and explain what had happened to an assembly of his fellow citizens at Yesler's Hall.

Unfortunately for the art of rhetoric, no account exists of just what it was that Mercer said that night. The *Puget Sound Daily* failed to report the gist of his remarks; it only noted that "marked attention was paid, the speaker being frequently applauded." Conspicuous in the audience were the Mercer girls themselves, "the fair immigrants who had so recently arrived." In the opinion of the *Daily* it was the obvious confidence these ladies placed in Mercer that vindicated him. The meeting closed "apparently with the best of good will towards Mr. Mercer and all concerned." Wisely deciding not to press his luck, Asa Mercer himself soon married one of the fair immigrants, Annie Stephens, and eventually ended up in Wyoming as a rancher.

Meanwhile the males of Seattle and the loggers from camps in the area were little better off than before. The male-female ratio in the Northwest was destined to be off-balance for a long time. John Pennell's solution —companionship for a price— was, in practical terms, the only one available. Pleasure houses proliferated in Seattle and Portland, the twin meccas of every logger because they had the biggest and most licentious Skid Roads in all the timber kingdom.

The greatest of them all probably was Duke Evans' Paris House in Portland, in size a bawdyhouse counterpart to Erickson's saloon just a few blocks away. Like Erickson's, the Paris House stretched the length of a city block. In the criblike rooms on its two floors were housed exactly one hundred girls—or so Duke Evans liked to claim, although the police, when they bestirred themselves, never netted more than 83 in any one raid. White girls held forth on the lower floor, girls of darker hue on the upper. Successive waves of moral indignation rolled over the Paris House but never submerged it; in fact, it seemed to thrive on the free publicity. Not until 1907, when Duke Evans retired, did the place close down. By then the rough, crude, Babylonian-sized bawdyhouse was going out of style anyway. Up in Seattle an entertainment genius named John Considine had perfected a more glamorous kind of pleasure palace. It combined the saloon, the theater and the house of ill fame all in one attractive package.

Considine did not invent the idea. Such places had existed before and were known as "box houses." There was a stage and a bar, and around the unoccupied

walls ranged a number of small, boxlike compartments in which patrons could watch the show, drink and make love. But before Considine came along they were usually basement dives, often flooded in the rainy season; the booze was usually bad and the women employees were Jills-of-all-trades, hustling drinks and singing badly on stage when not turning tricks in the side cubicles. Considine's idea was to tidy up the box house and to specialize: he hired performers who did nothing but sing and dance, and floozies who did nothing but entertain loggers and other spenders one at a time. For the opening of his People's Theater, he imported the celebrated dancer Little Egypt. Admission was only a dime. The fact that Little Egypt had been jailed for dancing in the nude at Sherry's in New York did not hurt her drawing power, and an observer reported that she did indeed perform "an extremely interesting dance" to "tremendous applause."

Every evening, the band from the People's Theater whooped up business outside the entrance and mu-

Given a rousing send-off from New York in 1866, the S.S. *Continental* sails for the West with a group of "Mercer Girls." These *Harper's Weekly* sketches offer a glamorized version of the voyage of the brides-to-be. Actually, they suffered from *mal de mer* and a diet of boiled beans.

sicians from a rival box house did likewise. The Seattle *Post-Intelligencer* described a typical encounter: "It is about 8 o'clock in the evening that the battle begins. The players of the brass band on the west side of the avenue file out from behind the swinging doors and, removing their coats, hats and collars, prepare for the fray. The selection ended, the leader of the orchestra lowers his cornet from his ruddy countenance, and bows low to the crowd surrounding him and his brave supporters.

"In the meantime, the champions of the opposition have taken their stations on a high platform built over the entrance of the People's Theater. There are three of them. The leader is armed with a violin. Another dark-faced young man strums a huge harp. The third of the challenged musicians defiantly pipes away through a husky clarinet.

"Suddenly around the block is heard the discordant blare of an untutored brass band and the voices of men and women upraised in a popular street ditty. But the words are strangely out of joint. They seem to have been adapted from a hymn book and misfitted to the tune. It is the Salvation Army! Fifty strong, the uniformed Soldiers of the Lord swing into the street in front of the theatre and march up toward their Yesler Way barracks, flags flying, torches smoking and sputtering musicians playing like mad." The box house bands are taken aback, but they blare out again, "and pandemonium reigns. The crowd cheers. The Salvationists are outpointed two to one in the contest, but on they march, happily unconscious of the fact, leaving the theatre band to finish that enlivening melody, 'There'll Be a Hot Time in the Old Town Tonight.'"

The battle of the bands having ended in a draw, the rival box houses' entrance halls quickly filled up as coins "by the quart" poured into the ticket sellers' desks. When the shows began inside, "gaily dressed girls and women, wearing abbreviated skirts, plenty of paste jewelry, and a superabundance of paint, powder and false hair, come and go without hindrance to all parts of the house, soliciting patrons to buy drinks, upon which they make a commission."

John Considine's formula was foolproof. He was so successful and made so much money that he went legitimate, took the theater part of his idea and parlayed it into the nation's first vaudeville circuit. He left the Skid

Road and its uncouth loggers for others to exploit.

More than one member of the fair sex had by now concluded that the money to be made on Skid Road could just as well flow into a lady's handbag as a gent's wallet. The madams who provided female companionship to the loggers, sailors and mill hands of Portland and Seattle were a colorful group. Liverpool Liz, born Elizabeth Smith, presumably in England, ran the fancy Senate Saloon on Portland's West Side. Though she was no beauty, Liz was always stunningly dressed, her bosom flaunting a mesmerizing necklace of diamonds and gold said to weigh four pounds. Liz disdained knockout drops and rolling; she liked to acquire a man's cash more subtly. A logger would walk into the Senate, find three or four others at the bar, and grandly say, "Come on, all you fellows, and have something on me!" Instantly a dozen unoccupied girls would trip downstairs from their cubicles to join in—the bartender having pressed a buzzer to summon them. As Liz well knew, no self-respecting logger would refuse to buy all the girls a drink too.

Like John Considine, the best of the madams knew how to put on a good show. Mary Cook, a 285-pound valkyrie who ran the Ivy Green in Portland, acted as her own greeter, standing at the entrance to the bar amiably waving a large cigar. If a man extended a finger, she would blow three perfect smoke rings around it and wish him luck. Big Mary was also her own bouncer. If need be, without for a moment losing her sunny disposition, she could pick up a troublemaker and throw him bodily out the door.

For sheer resourcefulness, even these clever ladies could not hold a candle to Nancy Boggs. Nancy was a rugged individualist who could not abide paying off the police. Her solution was a "whiskey scow," as Portlanders called it, which she moored in the Willamette River. It was a two-decker, 40-by-80-foot houseboat with a gaslit bar and dance floor on the lower deck, and living and working quarters for two dozen girls topside. A floating hellhole, cried the better elements in town, who tried to sic the police on Nancy. But Nancy had figured out how sweet were the uses of mobility. Since Portland and East Portland across the river had not yet merged and still had separate police departments, she could avoid either branch of the law simply by weighing anchor and heading for the opposite

bank of the Willamette. Once, when both police departments were after her simultaneously, Nancy simply let her craft float downstream beyond both city limits. Ultimately, after a number of her customers were lost overboard, she gave up the floating life, moved to dry land and paid her bribes like the rest.

It was in Seattle, where the Skid Road was born, that a lady named Lou helped write the Road's last flamboyant chapter. Lou Graham, as she called herself (she began as Dorothea Georgine Emile Ohben), settled in Seattle in the late 1880s. She set up a house of prostitution on Washington Street and, after the fire of 1889 destroyed 50 blocks, rebuilt her place on a grand scale. Seattle was bursting with growth, in the vice business no less than in the lumber trade. One reform committee claimed that there were 2,500 prostitutes in town. "Scores of women are known to police as thieves and pickpockets," wrote the *Telegraph*. "Robberies occur in broad daylight. Men have been known to go into alleys with hundreds of dollars and walk out half an hour later penniless."

Sensing which way the wind was blowing, Lou Graham made an agreement with the local government. She promised that nobody would ever be robbed in her four-story brick mansion, that it would be as genteel a place as any logger—or, for that matter, any alderman —ever entered, and that she would contribute to municipal funds a fine, or license, of $10 per month per girl and $50 per month per gambling table. Any city official could be her guest at no charge. In return she wanted an understanding that she would not be raided.

Before long, such levies from Lou Graham's and similar establishments were pouring $100,000 a year into the city treasury. To keep her face and her place in the public eye, small, dark-haired, blue-eyed Lou, bejeweled under a plumed hat, took her best-looking girls for Sunday rides through the business district. She was arrested only once, by a new cop who had not been briefed on the rules. Altogether, Lou was Seattle's most magnificent madam, and upon her death in 1903 she left about a quarter of a million dollars—to the county school system. To the county school system? When they heard about Lou's will, the loggers who had brawled, boozed and womanized along old Skid Road shook their heads admiringly. Truly, an era had ended—but Lou had seen to it that it ended in style.

Decorating the Hill Brothers' Pool Room in Mill City, Oregon, this full-blown dryad was typical of the loggers' ideal of feminine beauty. Nudes graced virtually every establishment that the woodsmen frequented.

6 | Big Business invades the forests

The timber that marched down the slopes to Puget Sound at the turn of the century appeared to continue right across the water, in a forest of masts. Lumber ships from all over the world crowded harbors such as Port Blakely and Port Gamble, testifying that the Northwest's lumber had become far more than a regional industry. But the lumber fleet was even then yielding its status as the prime mover of Northwest timber to a newfangled carrier: the railroad. The steel tracks that pushed their way west across the United States from the Great Lakes into Douglas fir territory in the 1880s and early 1890s did what the earlier transcontinental line farther south had not: they spurred Lake States timber barons to transfer their operations out to the Northwest. These highly industrialized lumbermen used the lines to carry lumber to the greatest market of all: the whole of the United States east of the Rockies.

Sharing the waters of Port Blakely, Washington, with a boom of unmilled logs, a flotilla of lumber ships from many countries waits to load.

Tycoons, Wobblies and reformers

In 1890, when young Tom Ripley arrived in Tacoma on a visit from Vermont, people from all over the lumber-hungry world were pouring into that booming town. "East and West are meeting in Tacoma," the Yale graduate wrote. "Chinese meet Philadelphians; Japanese meet Bostonians; native sons of California meet Down-Easters from the State o' Maine. Lumberjacks from the Saginaw, fresh from the skidroads and stepping high, meet Siwash Indians from their camps in the mudflats below the bluffs."

Ripley decided to cast his lot with the lumber industry; in time, he became president of the Wheeler-Osgood Company, one of the wealthiest woodworking firms in the area. Tacoma, which he called "the City of Destiny, the harbor of progress, the tree-studded land of plenty," richly fulfilled its promise.

As Ripley so clearly saw, hundreds of thousands of enterprising people moving West were powering the region's growth. Between 1880 and 1910, when the pioneer period was drawing to a close, the Pacific Coast's population nearly quadrupled, reaching 4.1 million. The area's lumber centers expanded at an even faster rate. Tacoma's population soared from a little over 1,000 to 83,000. Seattle's population, about 3,500 in 1880, skyrocketed to 43,000 by 1890, doubled again by the turn of the century, and more than redoubled in the next decade, topping 200,000.

Meanwhile, massive investments of Eastern capital transformed lumber from a thriving regional trade into a colossal, highly mechanized industry that served the entire burgeoning United States and much of the world

---

Frederick Weyerhaeuser, the greatest lumber baron of all, embarked upon Western logging in a style befitting the era: he bought 900,000 acres of timberland from the Northern Pacific Railroad for $5.4 million.

besides. In the same three decades, production from the Western forests soared from some 660 million board feet a year to more than eight billion.

But with the many efficiencies of huge size came a host of problems that had for years plagued industrialists back East. The Western timbermen were soon sorely tested by the beginnings of labor unrest among their mill hands and logging crews, and by crusading federal trust-busters who suspected them of price-fixing and other illegal practices. Furthermore, as the nation awakened to the rapid depletion of its abundant natural resources, lumbermen were fiercely attacked by conservation-minded reformers, who charged them with ravaging woodlands that belonged to all the people.

The industrial age was launched in the Western forests by the American lumbermen's long-time partners in growth, the railroaders. In the 1850s and '60s, thousands upon thousands of miles of new track had been laid as the railroads crisscrossed the eastern half of the nation. Each mile required 2,600 heavy wooden ties; the big Eastern lumber companies supplied them. But this enormous business was only the beginning of their profitable interchange. Wherever the railroads pushed their steel rails they brought new settlers and opened new markets for lumber. Then, in 1869, the Union Pacific Railroad and the Central Pacific met at Promontory, Utah, thus linking Sacramento to Omaha, Nebraska, and the East. Within the next decade crews were laying tracks northward into Oregon and Washington to open up the interior and give loggers there an overland freight route to San Francisco, which heretofore could be supplied only by sea.

Western lumbermen were grateful to railroaders for the customers they were bringing from the East, and for the north-south freight link. But they were slow to take advantage of an even greater boon: new access to the East, where the population was swelling with waves of immigrants, creating a need for more lumber than Eastern sources could supply. It was an understandable failure, however. Western loggers had grown up in isolation; their thinking had always been geared to local markets and to seaborne exports to Pacific ports. The growing output of the California mills continued to be used chiefly by the state's increasing population (864,000 by 1880) and little lumber was left to send East. Meanwhile, Washington and Oregon were capable of producing much more lumber than their populations (250,000 in 1880) could use. But Puget Sound and the heartland of the Northwestern forests had no direct access to the East.

That vital missing link was finally supplied in 1883, when track-laying crews for the Northern Pacific Railroad closed the last gap in a 1,700-mile route between Duluth, Minnesota, and Portland. A decade later, James J. Hill's Great Northern Railroad hooked up Minneapolis with Seattle. The Eastern network of roads tied into both Minnesota terminals. Now lumber from the world's richest forests took only a few days to reach the vast Eastern market.

The completion of the Northern Pacific had a galvanic effect on lumbermen—and not just the Northwesterners. The great timber barons of the Lake States had for decades been freighting lumber eastward by rail and were fully aware of that trade's tremendous value. They had long been interested in the forests of the Far West and the Rockies, and now the Northern Pacific made it feasible for them to buy timberland there. Furthermore, the railroad went through just when the Lake Staters needed new woodlands to conquer.

By 1880, the forests of Michigan, Wisconsin and Minnesota had been heavily logged for more than 40 years; they were fast reaching the level of depletion that had prompted lumbermen to abandon Maine. By contrast, the immense Northwestern forests were still virtually intact in spite of a half century of logging, and timberland prices there were a steal. Small property owners, who had claimed 160 free acres from the public domain, were willing to sell their land for a modest sum, or to lease stumpage rights for a pittance. Big landowners were offering substantial tracts for as little as $1.90 an acre, though they asked higher prices for very large holdings on the impeccable theory that these could be logged more economically and were thus

worth more. But even if one had to pay as much as $7.50 an acre, that sum was the lumber value of three giant Douglas firs, which usually grew 50 to the acre.

The final spike in the Northern Pacific roadbed had hardly been driven before the Lake Staters started sending their agents west to appraise the forests. Almost everywhere the timber cruisers saw rich acreage. But the land that attracted them most belonged to the railroads; the United States government, as part of its long-time policy of fostering the development of the country, had granted the roads millions upon millions of acres, mostly in a checkerboard pattern of 640-acre (one-square-mile) sections along their rights of way. Under their agreement with the government, the roads had to sell off sections to finance track-laying and operating costs and, being land-poor, they were desperately eager to do just that.

One of the first Midwesterners to head for the Pacific Coast was a Minnesotan named Chauncey Griggs. In 1887, Griggs formed a partnership with some like-minded lumbermen, moved to Tacoma and set up the lucrative St. Paul and Tacoma Lumber Company, logging and milling on 80,000 acres by the Northern Pacific's right of way. The firm was one of the first on the Pacific Coast to rail-freight Douglas fir lumber and western redcedar shingles to the East.

Singly or in affiliated groups, more and more Lake States companies packed up and moved west by rail. For a goodly number of Lake States barons, the move was the second stage in a general migration that had brought them from the New England pineries to the Midwest. Gary B. Peavey, of the Penobscot family credited with inventing the famed logging tool that bore their name, migrated first to Minnesota, then in the 1880s to Puget Sound. David Whitney, originally a Massachusetts man who logged extensively along Michigan's Saginaw River, bought great tracts of timber in Oregon, Washington and northern California—and presided over his woodland empire from his magnificent mansion in Detroit.

By the early 1890s, the transplanted Lake Staters were cutting a Bunyanesque swath in the Western lumber trade. Using their bulging bank accounts with shrewdness and vision, they built efficient new sawmills, installed high-speed saws in expanded old mills, connected the forests to the mills with private logging railroads and—by increasing their production steadily—drove unit costs down to a point where they could sell lumber at unbeatable prices.

These developments caused much anxiety—and some bitterness—among the local lumbermen whose firms had been working the California and Northwest forests for decades. They had spent a lifetime building their businesses only to see rich, pushy newcomers found larger firms almost overnight. The Westerners had done their best to keep abreast of technological improvements and had invented invaluable logging machines. But now, to stay competitive, they obviously had to expand their facilities and invest in new equipment on a scale out of all proportion to their finances. Most of them simply did not have the credit—much less the capital—to match the newcomers saw for saw, donkey for donkey, locomotive for locomotive.

In 1896, George Emerson, a manager of Asa Simpson's far-flung but old-fashioned Northwestern Lumber Company, warned his boss, "Without all these costly appurtenances, within a few years a mill plant will have fallen so far behind the march of events as to be out of the race." Unhappily, Emerson did not win his point; Simpson was loath to modernize, and before long his enterprise was reduced to second-rank status. A very few, notably Pope & Talbot, were prodded by the competition of the Lake Staters to expand. And they owed much of their increased income to a lesson they learned from the same Lake Staters: the value of rail trade with Eastern markets. In 1896, Oregon lumber, supplied by inexpensive rail freight to the East, brought an average of $11 per thousand board feet—half again as much as the local price. No wonder the volume of Northwestern lumber rail-freighted East shot up; in 1899, some 10,000 carloads of lumber were sent inland from the Pacific Northwest; two years later, the number of carloads approached 25,000.

By then, every lumberman on the West Coast realized that the glory days of the do-it-yourself owner-operator were gone forever. There was still room for the rare throwback like Simon Benson *(pages 198-201),* who had the entrepreneurial genius to make it on his own. But most of the old Western giants had died off, leaving their personal empires in ruins or in the hands of a new generation of organization men. And in this new corporate age of logging, it turned out that the biggest Western operator, the leader to whom everyone looked, was a Lake Stater, who ran literally dozens of companies in eight states through echelons of directors, presidents, managers, specialists, field bosses and assistants. For many years, his methods and policies had dominated the lumber trade east of the Rocky Mountains. And now, expanding westward at the head of a phalanx of other entrepreneurs, he shaped the lumber industry of the entire nation.

This colossus, this quintessential timber baron, was Frederick Weyerhaeuser, whose success story bore eloquent testimony to the basic truths in the American Dream: that hard work and native intelligence—plus a bit of luck—could build a man an empire in the New World. Weyerhaeuser began his long career in 1856 as a 21-year-old German immigrant, toiling for day wages in a sawmill at Rock Island, Illinois. Four years later, he bought that mill with his brother-in-law, Frederick Denkmann; and in the next two decades he brilliantly pyramided his profits into more mills and great tracts of Wisconsin timberland, creating one of the Lake States' largest complexes of companies. In the same period, Weyerhaeuser used his exceptional talents as an organizer to set up two large cooperatives. Under his skillful guidance as president, their forest-to-market service for member firms helped bring peace and order to the region's chaotic lumber business. These feats prompted a trade publication in 1884 to call him "the Master Spirit of the Mississippi River Valley as far as logs and lumber are concerned."

For all of his fortune and fame, Weyerhaeuser bore little resemblance to the stereotype of the ruthless, domineering, ostentatious tycoon. Quiet and cautious by nature, he clung to his simple life style; he went to bed early, to church regularly and to the kitchen often for buttermilk, which he credited with his good health. Far from seeking dictatorial power, Weyerhaeuser preferred to share his responsibilities—along with his profits and risks—with a large circle of associates. He owned only a minority interest, usually ranging from 15 to 20 per cent, in most of his companies. Yet his many partners habitually followed his lead because his decisions had proved so profitable. Knit together by his influence into an alliance that would last for three generations, they became known as the Weyerhaeuser Group. ◉

Simon Benson and his seagoing cigars

Even after logging submitted to domination by giant corporations, there was room for the lone enterpriser who set out from scratch to make it on his own. Simon Benson was one such bright and determined individualist.

Benson was born on a tiny farm at Gausdal, Norway, in 1851. As a lad, he was known as a dreamer; when he told his elders that some day he would be a great landowner, they said: "Too bad! He looks at the stars instead of the ground." But Benson was a star-gazer with uncommon perseverance.

Emigrating to America with his family at 15, Simon spent his early years learning logging in Wisconsin. When

Simon Benson

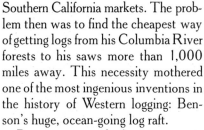

he was 27, he headed west to Oregon. Starting in a lumber camp at $40 a month, he soon saved enough to buy six oxen and set himself up as an independent bull whacker. His profits enabled him to buy a little land and start logging for himself.

In 10 years, Benson had crews working 4,000 acres near Cathlamet, Washington. He bought rights to the best timber and kept up with technical developments, getting rid of his ox teams, for example, when he saw that donkey engines could cut costs in half.

More than once he went broke as the result of fires or depressions, but he always bounced back. By the turn of the century, Benson ran no fewer than 15 camps and 75 miles of logging railroads along the Columbia. He calculated his profits at $2,800 a day. But there was, he felt, still room to grow.

In 1906, Benson built a large sawmill in San Diego to serve the growing

Southern California markets. The problem then was to find the cheapest way of getting logs from his Columbia River forests to his saws more than 1,000 miles away. This necessity mothered one of the most ingenious inventions in the history of Western logging: Benson's huge, ocean-going log raft.

Benson was not the first to try transporting logs by sea, tied together in huge rafts. However, earlier attempts on both coasts failed when heavy seas twisted and then broke apart the massive rafts. Benson built his raft in such a way that it was almost as rigid and streamlined as the hull of a ship.

To shape his long, pointed "hull," he devised a cradle that acted as a mold. When half the logs were hoisted into the cradle, a heavy anchor chain was stretched from stem to stern to serve as a backbone for the raft. After the rest of the logs were in the cradle, more chains—tons of them—were fastened like belts around the logs to clasp the resulting raft together. The cradle was then opened and the raft floated out.

In 1906 the first Benson raft completed its journey to San Diego without losing a log. It was followed by scores more—some of them 1,000 feet long, carrying six million board feet of timber. The rafts provided Benson's mill with an endless supply of timber and helped make him a multimillionaire.

Someone once asked Benson if he had ever written back to Norway about how he had fared in the New World. "No," Benson replied, "they would have tapped their heads, smiled, and said, 'Still building castles in the air.'"

The wooden mold or cradle in which a seagoing Benson log raft will be assembled rests in a quiet backwater of the Columbia River. The cradle was built in several long sections that were afterward separated vertically to release the completed raft.

Loops of heavy chains, fastened together by massive turnbuckles, secure the precisely aligned logs in a Benson raft. It sometimes required 250 tons of chains to maintain the raft in a properly rigid mass.

A cigar-shaped Benson raft nears completion, its cradle, half of which is attached to pilings on the near side, almost submerged under the weight of the logs. The derrick was used to lower logs and affix chains.

Midway through launching, a raft has been pulled free from half of its cradle, which is now riding high in the water moored against the pilings at right. Assembling a raft could take as long as seven weeks.

Aided by river tugs, a Benson raft passes Astoria, Oregon, on the first leg of its two-week trip to San Diego. Near the mouth of the Columbia, a sea-going tug took over for the 1,100-mile ocean journey.

eyerhaeuser started moving west around 1890. As the Wisconsin forests dwindled and loggers pushed on into Minnesota, the great baron stayed close to the center of industry by shifting his headquarters in 1891 from Rock Island, Illinois, to St. Paul. But even as he set up operations in St. Paul, Weyerhaeuser understood that the Minnesota forests would soon be waning, and he began systematic efforts to locate timberland for large-scale operations outside the Midwest. Acting as his own timber cruiser, he made two inspection trips to the pine forests of Louisiana and adjacent states, but declined the prospect of a huge tract because he thought the $2.5 million price was too high. He visited the California redwood lands but decided against buying because the wood was so little known in the East. He even traveled to Alaska, though transportation costs from that territory priced its lumber out of the market. And of course he made several trips to inspect the immense Douglas fir forests of Washington and Oregon.

Through the 1890s, Weyerhaeuser's growing interest in Northwestern timber was encouraged by his friend and neighbor in St. Paul, railroad magnate James J. Hill. When Weyerhaeuser bought his new home at 266 Summit Avenue in 1891, Hill's Great Northern Railroad was still building toward its western terminus

Redwood lumber lies drying at the new "upper mill" of the Kings River Lumber Company, at Millwood, California. Shortly after it opened, the

at Seattle. Since Hill was the only railroad tycoon to finance a transcontinental line without government land grants, he had little timberland to sell; he was after Weyerhaeuser mainly for the lumber-freight business he might provide. But by the late 1890s, a few years after the Great Northern reached Seattle, Hill had gained control of the foundering Northern Pacific Railroad, which did own government-granted land—an incredible 44 million acres, much of it in prime timber country. Then Hill was supremely anxious to sell Weyerhaeuser those wonderful forests.

In the autumn of 1899, Hill lent Weyerhaeuser his private "palace car" for another trip west, and supplied him with a good shepherd: William Phipps, commissioner of the Northern Pacific land-sales office. By the ides of November, Weyerhaeuser and a number of associates were rolling west in baronial style toward Tacoma. His retinue included Rudolph, the third of his four sons; his partner, Frederick Denkmann; A. E. MacCartney of his St. Paul law firm; Joseph Lockey, cashier of the National German-American Bank in St. Paul; and eight logging associates from three states. The party arrived in Tacoma on November 17 and found an excited reception committee at the station.

A reporter for the Tacoma *Evening News* earnestly asked Mr. Weyerhaeuser about the purpose of

upper mill raced the company's lower mill to see which could produce more board feet in a day. The upper mill won, 76,000 to 72,000.

his trip. Replying in his usual pleasant, professorial way, the timber baron admitted that he had "several good propositions in view. You have plenty of fine timber in Washington, and so has your southern neighbor, Oregon. We are just looking around on this trip. We may buy a great deal of timberland here and we may start up several mills. We'll see what can be done."

Then Weyerhaeuser uttered a cautionary note: "There is one disadvantage on the Coast. This is the railroad rate for shipment East. We are asking the railroads to give us a reduction of rates." That somber sally was for the benefit of Jim Hill, who had already set a bargain rate of 40 cents per hundred weight of lumber shipped from the Northwest to St. Paul. Weyerhaeuser knew that Hill could not reduce the rate any further; his railroads often had to bring empty cars west just to freight lumber east. But Weyerhaeuser wanted leverage to bargain for railroad land.

Next morning, the Weyerhaeuser party boarded Hill's car for a tour of timberland in three counties. A few days later, they returned to St. Paul, leaving Tacomans to wonder what the inscrutable lumber magnate had on his mind.

For six weeks not a word on the subject issued from Weyerhaeuser's St. Paul office. Then, on January 3, 1900, the great man gave his answer. In one of the biggest real estate transactions in history, the Weyerhaeuser Group announced the decision to purchase no less than 900,000 acres (1,406 square miles) of Northern Pacific timberland, mostly Douglas fir, in Washington. The price was $6 an acre. To swing the $5.4 million deal, Weyerhaeuser and Denkmann put up $1.8 million, while their associates, 15 of them, put up the remaining $3.6 million.

The next step was to form a company to control the domain. Organization meetings produced a flurry of debate that shed much light on Weyerhaeuser's character. His partners proposed calling the firm The Weyerhaeuser Timber Company. Weyerhaeuser demurred; he was only one of a number of investors. But the partners would hear none of it. The Weyerhaeuser Timber Company it was named — with Frederick Weyerhaeuser as its president.

The general manager chosen to superintend the new company was a prime example of Weyerhaeuser's uncanny ability to find the best man for every key job.

He was George Smith Long, a lanky, Lincolnesque Hoosier who had earned Weyerhaeuser's respect while working for a rival firm in the Lake States. Long was a top judge of lumber and an authority on marketing. Weyerhaeuser had to bid high to woo Long away, but finally signed him for the then handsome salary of $5,000 a year and credit to purchase a $100,000 block of stock in the new company.

Weyerhaeuser's operations in the Northwest were at first inhibited by the sheer size of the land purchase. Conservative members of the board of directors considered it imprudent to expand rapidly because of the energetic federal trust-busting activities that had broken up John D. Rockefeller's oil empire. Though not even the Weyerhaeuser Timber Company was powerful enough to dominate the lumber business in the Northwest, muckraking journalists greeted the big timberland deal with vague allegations of a "lumber trust"; it was well, the directors thought, to keep a low profile. Weyerhaeuser held back for another reason: he was genuinely reluctant to hurt small owners and wanted to give them time to adjust to the presence of his corporate giant. Thus, when George Long got his orders in May 1900, the directors authorized him to buy more timberland at favorable prices, but they warned him to buy or build mills sparingly and when possible to use local contract loggers to supply wood.

Long went to work with consummate skill and tact. With Weyerhaeuser's approval, he built a new mill at Grays Harbor and another near Portland. He also shelled out $240,000 for a small old mill in the town of Everett, and with new machinery jumped its output from 28 million board feet in 1902 to 39 million three years later. With Long's vigorous land-buying program, Weyerhaeuser Timber's lands grew to 1.5 million acres by 1905. But more than half the logs sawed in the company's Washington and Oregon mills came from other people's land. Whenever the price was low, Long supplied the mills by buying timber rights to huge tracts, shrewdly saving the bulk of the company's own forests as a reserve for the future.

Idaho, too, was beginning to contribute to the company's growth. In 1903, Weyerhaeuser and his associates established the subsidiary Potlatch Lumber Company with $3 million in capital and thousands of acres of rich white pine forest in the northern part of the state. Originally they planned to use existing mills to saw their trees into lumber. But they had barely started operations before they realized that the small local mills could not handle such a large operation. To choose a location for a new and much bigger mill, the directors, including Weyerhaeuser, arranged to meet in the frontier town of Moscow with William Deary, the rough-and-tumble veteran whom they had appointed Potlatch manager.

The meeting, held on the second floor of a general store, got underway before Deary arrived. "It was March," a company official reported, "and those Idaho springs were bad. Mr. Deary came in from the woods. He pulled off his boots and was warming and drying his feet and stockings beside the stove when he heard the group at the table decide to put the mill in Moscow. He jumped up barefooted, went to the table where the map was spread out, picked up a lead pencil and said in his thick Irish brogue: 'Gintilmen, there isn't enough water in Moscow to baptize a bastard! The mill'll go here.' He punched a hole in the big map over the Moscow Mountain on the Palouse River."

Deary, being a properly astute Weyerhaeuser manager, had chosen the ideal site for the mill and the company town of Potlatch. Within three years, the big new plant—connected to the forests with 45 miles of railroad track—had begun sawing 135 million board feet a year for a lumberyard that covered 65 acres. ◉

Douglas fir columns line "the world's biggest cabin," the 1905 Lewis and Clark Exposition's Forestry Building in Portland, Oregon.

A Victorian extravaganza that redwood built

Rare was the lumber baron who failed to build himself a mansion, and some of the homes were grand indeed. But for sheer, flamboyant opulence, none could match the redwood castle erected by William Carson at Eureka, on northern California's Humboldt Bay.

Carson arrived in California in 1850 to join the quest for gold, and stayed to log the redwoods. In partnership with John Dolbeer, of donkey engine fame, he built a huge logging and milling en-terprise. And in 1884, a multimillion-aire, he felt the time had come to con-struct a proper home for himself, his wife Sarah and their children.

The structure was two years abuild-ing, and when finished it was a triumph of Victorian elegance. Redwood was the primary material, inside and out, but the interior also boasted other woods — notably primavera, a prized South American hardwood. Lavish examples of the wood carver's art decorated the gables and many of the 18 main rooms. Carson imported French tapestry for the dining room chairs and Mexican onyx for the fireplaces.

The lumber baron lived contentedly in his castle for 26 years, until his death in 1912. Sometime during that period he repainted it in two tones, making it even more flamboyant than before. His house, as one writer put it, was "a man-sion to bestir Archimedes, who ex-claimed 'Eureka' (I have found it)."

WILLIAM CARSON

SARAH CARSON

Resplendent in its red-and-white striped roof, Carson's mansion, shown here shortly after its completion in 1886, was built by the lumber magnate's own mill workers, who were facing depression-year layoffs.

The roof is toned down but everything else livelier in the colors Carson later chose. The 68-foot tower overlooked his mill.

Bristling with lattice-work and carving, a third-floor gable juts from the house. There were eight gables, no one like any other.

Carson's interwoven initials, over each of the two main entrances, form part of the wrought-iron second-story-deck railing.

Huge carved supporting pillars stand on equally sturdy balustrades along the wide porch that skirts two sides of the house.

207

South American primavera wood, used for the front entry, surrounds stained-glass door panels depicting an ancient knight and lady.

Redwood and primavera infuse the hallway with richly contrasting tones. The chandelier is alabaster and the English clock is oak.

Paneled with Philippine mahogany and local redwood, the music room's double-flued fireplace frames a stained-glass window.

Mostly of oak, the dining room was modeled after one in Mexico City's Chapultepec Castle, which Carson had visited. Beveled mirrors and other glass are from England.

In the mansion's most elaborate stained glass, medieval figures representing science and the arts adorn the grand staircase.

Intricate wood carving, shown against red draperies, surrounds a music-room alcove. The organ was custom-made for Carson.

Moorish arches in the sumptuous second-floor hallway appear to exhibit still more wood carving, but actually the designs are molded plaster. The gleaming pillars at right are made of redwood, the railings of honey-colored primavera.

While Weyerhaeuser flourished from the moment he moved West, he and his colleagues had to deal with a number of acute problems. The catastrophic fires of September 1902 that set the forests ablaze from northern Washington to central Oregon burned over 20,000 acres of Weyerhaeuser's best timber. But no sooner were the fires out than Weyerhaeuser and General Manager Long launched an all-out effort to salvage whatever timber they could. Since scorched wood tends to harden and deteriorate rapidly, making it all but impossible to saw properly, they highballed the logs to the nearest mills. For the logging gangs, it meant long days of dirty, grinding —and hazardous—toil. In the end, they managed to save much of the timber.

Fires were something every lumberman understood; Weyerhaeuser expected them and was prepared to deal with them. But as the decade wore on, a new and extremely serious problem arose in the woods: wide-scale labor unrest among the loggers and mill hands. It was something that neither Weyerhaeuser nor his fellow lumber barons had met before, and it baffled them.

Though Weyerhaeuser and the others had experienced occasional minor grumbling among the crews back in the Midwest, relations with their men had been generally good according to the standards of the times. But the times had changed. In the age of big business, loggers and mill hands seldom dealt directly with their top bosses, and they found it increasingly hard to win raises, promotions or redress for grievances. The owners' standards did not change nearly as fast as the times: wages remained low, risks high, hours long and working conditions poor. The men began listening to labor organizers, and started banding together for protest demonstrations and even strikes.

Weyerhaeuser and his colleagues may not even have noticed it at the time, but the year 1905 saw the formation of a new, all-encompassing labor union whose trumpet blast to action—"Workers of the World, Unite!"—would soon win many followers among the loggers and mill hands of the Northwest. The union was the Industrial Workers of the World, known as the I.W.W.; its members were referred to by a name that was never satisfactorily explained: Wobblies. (Legend has it that the name originated with a Chinese restaurant owner in Vancouver who befriended the unionists and fed them on credit. Pronouncing I.W.W. was too much for this Oriental gentleman; it came out sounding vaguely like "wobbly," and so it remained.) The I.W.W. was led by Big Bill Haywood, a huge one-eyed miner from Salt Lake City who had earned the enmity of big business as president of the radical Western Federation of Miners. Haywood's goal was magnetic: he simply wanted to organize every working stiff everywhere into one invincible union.

By 1907, the Wobblies were seizing labor leadership from less effective union organizers throughout the Northwestern lumber business. That year, they tested their strength in a Portland sawmill strike, demanding an eight-hour day (down from 11 hours) and wage increases based on a $3 daily minimum (up from little more than $1.50). The strike was doomed to failure. But the very fact that loggers had summoned the courage to strike and the publicity their action generated won still more followers for the I.W.W.

Two years later, the Wobblies mounted a mammoth organizing drive in and around Spokane. It was a spectacular show and virtually tied up the area for weeks. Union speakers poured into town, delivered their spiel in the streets, and attracted so many crowds and converts that the authorities passed an ordinance making street oratory a jail-worthy misdemeanor.

The Wobblies were delighted; they had been handed a dramatic cause — free speech. One by one, some 600 Wobblies mounted the soapbox, got arrested and then proceeded to eat up the town's prison funds. When the town fathers realized it was costing them thousands each week to feed their prisoners, they restored free speech and released the freeloaders.

The lumbermen, Weyerhaeuser among them, were worried by the strength of the Wobblies. But they were not the sort to be intimidated by a bunch of "radicals." Though most of these corporate chiefs had started at the bottom, they had flourished under the free enterprise system; they cherished its ideal of untrammeled business, in which a boss was boss and the working man knew his place — until he, too, became a boss.

The lumbermen refused to recognize any union, and made only minor concessions to employees. Some gave workers a 10-hour day, small pay raises and insurance against on-the-job accidents. Though labor unrest increased, Weyerhaeuser's son, Frederick Jr., summed up

the lumbermen's view: "I hope that we shall not have to concede an eight-hour day, and I particularly hope that we shall not have to recognize the unions." In time, the lumbermen had to do both.

By 1910, the Weyerhaeuser Timber Company had amassed an impressive but paradoxical record. Though it owned nearly two million acres of timberland, more by far than any other company, the lumber it sawed that year was only a fraction of the total output of the Northwest. To acquire that immense domain, the Weyerhaeuser stockholders had put up $12.5 million. But they had received not a single penny in dividends; all profits had gone back into the business to buy new lands and more efficient machines. It bothered the old baron not one whit that some stockholders were pressing for a payout. He had built solidly for the future. The timber was there; it would take but one order to make the Weyerhaeuser company the biggest producer in the world. In the Northwest and throughout the nation, the operations of the Weyerhaeuser Group were in the hands of talented, vigorous executives, experienced at working on their own with little intervention from the partners and directors. The corporative boat would not be rocked when Weyerhaeuser died.

In his declining years, Weyerhaeuser still heard the muckrakers' charge that because of his huge holdings and vast influence he exercised some sort of monopoly in the lumber business. He shrugged that off. But he took more seriously another familiar accusation — that because of the lumbermen's wasteful pursuit of their trade, the country was threatened with a timber famine. Weyerhaeuser knew that beneath the ill-informed attacks the problem was very real. He did not believe that the new plans for conservation were economically practicable for lumbermen, but he deplored irresponsible logging as much as anyone. All too often, he had seen loggers waste timber, cutting trees with unnecessarily high stumps, leaving logs to rot in the woods and to serve as kindling for fires. To Weyerhaeuser, as a prudent businessman, it was only sound practice to minimize waste in order to maximize profits. This view and his deep concern animated his last words. As he lay dying in Pasadena at the age of 80, the great timber baron muttered, "Cut 'em low, boys, cut 'em low."

The idea of conservation — with restrictions on the use of timber, water and minerals — was probably the

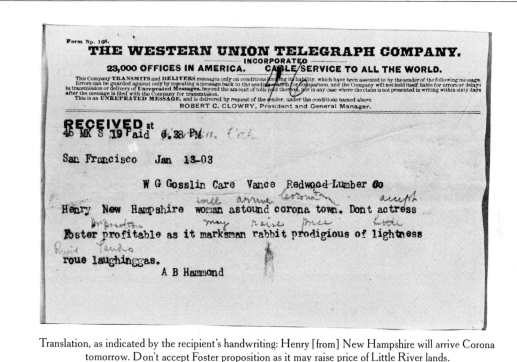

Translation, as indicated by the recipient's handwriting: Henry [from] New Hampshire will arrive Corona tomorrow. Don't accept Foster proposition as it may raise price of Little River lands.

most un-American idea that Americans had to face in their migration westward. After all, the nation had been born amid nature's abundance and had been raised to take it for granted. It was natural to hurry west to develop those resources, not to ration them. It was natural, too, for Americans to pooh-pooh the idea that they had been profligate with their resources—particularly with their seemingly endless timberlands.

Reformers spent most of the 19th Century persuading the government that forest conservation was an urgent need, and it took even longer for Congress and the Presidents to learn that hasty, well-intentioned laws would not make it work. Benefits would come only with patience and the cooperation of everyone.

The need for conservation had been recognized and duly announced as far back as colonial times. In 1789, a spokesman for the new U.S. Navy warned the Senate that the oak forests along the Atlantic seaboard should be husbanded carefully, since it took some 2,000 oak trees—about 57 acres' worth—to build a single large man-of-war. To protect the vital resources, Congress passed four forest-reserve acts between 1800

and 1827. But immediately a problem arose that would haunt officials for a century: many citizens kept right on helping themselves to the protected trees.

In 1846, Solon Robinson, an eminent agriculture expert, wrote, "If by some means the public mind of America cannot be induced to preserve and cultivate forest trees, the day is not far distant when we shall be as destitute of timber as many parts of Europe." It was much too early for any such warning; lumbermen were still unconcernedly stripping the forests, and officials who tried to stop them only exacerbated their resentment of government intervention in their affairs. In 1851, the General Land Office, a branch of the newly established Department of the Interior, sent enforcement officers—referred to as timber agents—around the country to stop trespassers from felling trees on public land. In no time, the agents were denounced in Congress for "unmitigated oppression." Cries of outrage reached such a crescendo that the Land Office had to call off its agents. The people, insisted the lumbermen's champions in Congress, had urgent need of lumber —and what the people needed the people would get. ◉

The General Noble's ignoble fate

Fallers link hands across the base of the General Noble tree before felling it. The redwood, whose circumference of 63 feet flared to more than 90 feet at the base, was probably 3,000 years old at the time.

In 1893, when the World's Columbian Exposition was held in Chicago, many people were still skeptical about the size of the California redwoods, so believe-it-or-not displays were common in museums and at fairs. For the Chicago show, the federal government came up with a suitably grand idea. A 300-foot giant sequoia, outside the protective limits of the newly created Sequoia National Park, was chosen to contribute its trunk to the show. Ironically, it had previously been named the General Noble, for General John Willcock Noble, Secretary of the Interior and an advocate of forest preservation.

A crew of audacious fallers on scaffolding made a cut 50 feet above the ground and managed to drop the upper 250 feet without killing themselves. They hollowed out the top 14 feet of their 50-foot stump, leaving a shell the thickness of the bark — two feet — plus six inches of wood. After cutting this shell into staves, they sawed off two feet of the still-rooted trunk, creating a gigantic wafer 20 feet in diameter.

They then repeated the hollowing and slicing operation on the next 14 feet of trunk. When they were finished they had two disassembled cylinders 14 feet tall and a solid, two-foot-thick disk. That left a 20-foot stump, at once magnificent and pathetic, which local residents called the Chicago Stump.

In Chicago, the General Noble's parts were reassembled into a roofed observation tower. The visitors who climbed the 30-foot spiral staircase inside it passed midway through a hole in the two-foot-thick "landing" that was the solid section. The idea of the landing was to prove beyond a doubt that the exhibit had once been a real tree.

The jagged upper trunk rests amid the debris of rigging and scaffolding. As it fell, it kicked back against the stump, smashing the scaffold and forcing the fallers to leap to the top of the wildly vibrating stump.

Leaning on their tools, workers rest in the partially hollowed-out stump. As the vertical staves were cut, each was numbered and crated for shipping out of the woods by mule team and to Chicago by railroad.

The finished redwood observatory stands on the Mall in Washington, D.C., after the 1893 Chicago fair. It remained there for many years, until it was disassembled for storage—just where was later forgotten.

215

The American people did need wood, lots of it. During the 1850s wood constituted the bulk of the fuel the nation consumed. But the people were also wasteful. In fact, farmers who went west cut down more trees to clear their fields than loggers felled to sell as lumber; and they burned all their logs in a single great bonfire just to be rid of them. The worst of it was that many Western homesteaders, like colonial farmers in Maine, had destroyed their woods only to find that the soil was too thin and rocky for any crop but trees. Thus, they were losers twice over.

At the very least, a law was needed that would prevent westering farmers from claiming timber-rich lands unsuited to farming. Such a law was proposed in 1877 by Secretary of the Interior Carl Schurz, the first high official to campaign wholeheartedly for forest conservation. Schurz got nowhere. Congress was leery of imposing hard-and-fast restrictions, especially since they were virtually unenforceable. Most congressmen felt they had gone far enough by passing the Timber Culture Act of 1873; under its provisions a homesteader could double his claim from 160 to 320 acres of free land if—as a conservation measure—he agreed to grow trees on 40 acres. But that act backfired: settlers would claim an extra 160 acres of land that nature had already wooded and sell 120 acres of timber—while technically complying with the law by keeping 40 acres of trees. All of which simply convinced the legislators that the most realistic course was to legalize the cutting of trees that people would take anyway.

Mining companies, which needed beams to shore up their tunnels, were authorized by the Timber Cutting Act of 1878 to help themselves, at no cost, to trees from adjacent public lands. That same year, Congress passed the Timber and Stone Act, which provided that any citizen of Washington, Oregon, California and Nevada could buy—for his personal use only—160 acres of public timberland at $2.50 an acre, a price that grew more and more attractive as land values increased. Many lumbermen, true sons of their none-too-scrupulous age, soon devised a way to circumvent the law. Their gimmick was similar to the trick used to amass homestead land in the 1860s. Called the "dummy entryman" system, it featured a stooge who would, for a small fee (about $50), claim 160 wooded acres, swear that he intended to keep the land, then imme-

diately hand the title to the lumber company, which would log it in his name. No one knows exactly how many acres passed into the lumbermen's hands in this way, but the total was certainly enormous.

As late as the 1890s, when Congressmen were starting to heed the conservation gospel, their land laws still contained compromises that were to the advantage of lumber interests. One well-intentioned concession was the "land-lieu" clause of an 1897 act; under its provisions, anyone who wished to donate his woodland acreage to a protected forest area could claim an equal amount of land elsewhere. The clause was a windfall for lumbermen, who traded cut-over timberland for virgin acreage. It was an even greater bonanza for the railroads. The Northern Pacific cheerfully relinquished the least desirable 540,000 of its millions of acres in exchange for 320,000 acres of ponderosa pine in Oregon, 120,000 acres of white and ponderosa pine in Idaho, and 100,000 acres in Washington, most of it in Douglas fir. Strictly by accident, the deal was not a total loss for the people; returned to them was an invaluable heritage: Mount Rainier.

These land-grabbing opportunities persisted for years. Timber magnates kept a corps of highly paid lobbyists in Washington to influence legislators and bureaucrats—who were reluctant to step too hard on an industry that supplied the nation's ever-growing demand for lumber. Bribery and influence peddling were rife. Officials winked at odious deals; known malefactors went unpunished. But pressure was building. When the lid blew off in 1903, the shock waves reached all the way into the United States Senate.

The man responsible for the explosion was a crooked real estate operator, Stephen A. Douglas Puter. He was a master of the dummy entryman system, but he tripped himself up by being too greedy. Rather than pay a stooge to claim land, he registered land claims under the names of nonexistent dummies. Eventually he was trapped by a federal prosecutor named Francis J. Heney. To save his neck, Puter turned state's evidence; he testified that he and other lumbermen had bribed John H. Mitchell, a distinguished U.S. Senator from Oregon, to intercede with the Land Office commissioner to approve their claims. Mitchell defended himself vigorously in the Senate, and the Oregon legislature gave him a vote of confidence. But Prosecutor

216

Heney was unmoved. He saw to it that Mitchell was indicted, tried, convicted and sentenced to serve six months in prison—an indignity Mitchell escaped by dying of complications following a dental operation. When the results of the scandal were totted up, 34 men in Oregon alone had been sent to jail for stealing or helping to steal public lands.

There was plenty of muck to be raked in these sordid trials, and the sensation-mongering newspapers of the day raked it profitably. But muckrakers also performed the service of convincing Americans that something was rotten in the woodlands. Unfortunately, the public notion of conservation was limited. Most citizens thought it consisted mainly of stopping lumbermen from stealing public land. Actually, theft was a legal matter that had nothing to do with the larger questions of forest policy. The real problem was simple and unsensational: waste—profligate waste by careless or uncaring loggers, by unthrifty sawmill techniques, by spoiled consumers, by preventable forest fires.

It took years for the conservationists who understood these things to get their message across. At first, they were a small, powerless group of scientists and professional forest managers who called themselves foresters. The best-known member of the group was old John Muir, the lyrical naturalist who tramped about the wilderness extolling its spiritual values and damning the "laborious vandals" who cut down California's invaluable sequoias. It was Muir who told the world that lumbermen were continuing to fell trees in two Sierra groves even after the area was protected by law.

But the conservationists were poorly organized; and their most important works—facts and figures compiled by bookish professors detailing the wasteful havoc in the woodlands—were so technical that they were read by scarcely anyone save fellow conservationists. Yet as time passed and the ravage of the forests became increasingly apparent, people began to take the conservationists more seriously. The Sierra Club, which Muir founded in 1892 to fight forest mismanagement, began to win converts; by the turn of the century it counted 700 select members, whose influence far outweighed their numbers. But truly effective conservation had to wait until the right man came along. And as things worked out, the right man was a well-connected young zealot named Gifford Pinchot, who arrived on the scene just as the country's attention began to shift from exploiting new gains to securing old ones.

Even as a student at Yale in 1888, Pinchot had given a schoolmate of his—the same Tom Ripley who went into the lumber business in Tacoma—an early glimpse of the fierce determination that would fuel his forest-saving crusade. Pinchot, Ripley and some friends were discussing their hopes and plans after graduation. "Alone among us," Ripley later wrote, "Gifford Pinchot knew what he wanted. 'I'm going to be a forester,' he said. We laughed. 'What the hell's a forester?' we asked. Gifford drew us a picture of a young knight of conservation, rambling through the forest and telling the lumber baron how to spare that tree and log it off at the same time by something he called 'selective logging.' Gifford alone among us had the Earnest Purpose."

From Yale, Pinchot went to France to study at the National School of Forestry; there was no such school in America. He soon learned that all the techniques of forest management had long been in practice in much of Europe; woodlands once "logged out" had been restored and kept productive for centuries. Pinchot was particularly impressed by the French law requiring owners to replant completely denuded forest.

Pinchot returned home in 1891 and wangled a job running an experimental forest at Biltmore, the Asheville, North Carolina, estate of railroad magnate George W. Vanderbilt. He carefully put into effect what he had learned in Europe. His staff kept the forest floor clear of debris to reduce the risk of fire. Instead of clear-cutting, he selectively felled mature trees only, each according to the optimum size of its species. The remaining trees reseeded themselves, so that the forest was in a constant state of rejuvenation. Moreover, under controlled conditions, trees grew larger and faster than they did in the natural state.

Pinchot's early experience at Biltmore whetted his appetite for bigger things. In 1896, at the age of 31, he won an appointment to a seven-man national forestry committee, newly established as an advisory body under the Department of the Interior, and there he began a career as the nation's most effective champion of forest conservation.

Shortly after Pinchot's appointment, a majority of the committee members made a recommendation he regarded with misgivings—on the ground that it was too

What was once a fine stand of sugar pine lies in a jumble of logs and splinters after a "clear-cutting" operation whereby loggers simply felled every tree in a tract. Spurred by scenes like this, the United States expanded nearby Yosemite National Park in order to protect adjoining sugar pine forests.

Gifford Pinchot, who became head of the fledgling U.S. Forest Service, was an avid but diplomatic conservationist. He helped lumbermen get the right to log forest reserves, then urged them to cut selectively.

harsh on the lumber industry. The committee urged an extension of the controversial Congressional act of 1891 establishing several forest reserves, mostly in California. Like the earlier National Parks, the reserves were to be held inviolate; no timber could be cut, no grass grazed, no coal dug, no road built, no dam built to prevent flooding or to create a reservoir or to harness water power. The reserves were simply "locked up," made unavailable for any use.

In February 1897, just before President Grover Cleveland left office, he complied with the committee's suggestion, transferring some 21 million acres of timberland from the public domain into 13 protected forest reserves. Instantly, loud protests were heard from Western loggers and millmen. To them, the creation of reserves was a discriminatory measure that could

—if carried to extremes—put them out of business.

Pinchot sympathized with the lumbermen. He also realized, being a pragmatist, that his plans for practical conservation could not be forced down the lumbermen's throats, that only their cooperation would make the reforms work. So Pinchot started lobbying to temper restrictions on the forest reserves with the magic word "use." By the end of the year, Congress redefined the purpose of forest reserves: lumbermen could use the reserves, but under strictly controlled conditions. Decreed Congress: "No public forest shall be established except to improve and protect the forest within the reservations . . . and to furnish a continuous supply of timber for the use and necessities of the citizens of the United States."

Western lumbermen never really accepted the reserves, which quickly grew to 128 million acres. Nevertheless, they were soon making a profit on timber they cut selectively from the reserves, and the government benefited from their payments for stumpage rights.

In 1898, early in the administration of President William McKinley, Pinchot was appointed Chief of the Division of Forestry, which had been established in 1881 and was responsible for the forest work done by the Agriculture Department. He took the post with the understanding that he would be allowed to run his division as he saw fit, and that meant total reorganization. He enlarged his staff from 10 to 180 and his annual budget from $30,000 to $200,000. He attracted a group of brilliant young disciples, who gladly accepted wages as low as $50 a month in exchange for the honor of working with "G.P.," as they called him. His division offered to supply free technical advice to loggers, lumbermen and local governments, and quickly received 123 requests from 35 states. In 1900 Pinchot played a leading role in the establishment of a forestry school at Yale; that same year he and seven colleagues organized the Society of American Foresters, a professional association whose ranks quickly swelled with graduates from Yale and the new forestry schools at Cornell and Biltmore. Weekly meetings were held in Pinchot's Washington, D.C., home.

About that time, Pinchot launched still another campaign—this one to improve efficiency by transferring all government forest work to his Forestry Division. His special target was the General Land Office of the In-

terior Department; Pinchot felt that the Land Office was more interested in catering to lumbermen's desires than to the needs of the nation as a whole, and he was determined that his foresters in the Agriculture Department would manage the country's timber lands. Presently, Pinchot got a powerful ally for his drive: President McKinley was succeeded in office by G.P.'s friend T.R.—Theodore Roosevelt.

Coming from wealthy families that traveled in the same circles, Pinchot and Roosevelt had known each other for years and had developed a mutual liking and respect. Both men were nature lovers, conservationists and crusaders. Both had camped under the Western stars with John Muir. Both were sports enthusiasts and physical-fitness addicts. Once, in 1899 when Roosevelt was New York governor and Pinchot was visiting him in Albany, T.R. playfully challenged Pinchot to a wrestling match and pinned him to the floor. Pinchot suggested that they switch to boxing gloves—and quickly knocked down T.R., who guffawed at the whole thing and thought it "bully."

Roosevelt and Pinchot saw eye to eye on the need to husband the nation's timber—and more important, how to do the job. The natural resources section of the President's first message to Congress stressed Pinchot's theme of proper use: "The fundamental idea of forestry is the perpetuation of the forests by use. Forest protection is not an end to itself; it is a means to increase and sustain the resources of the country."

However, Pinchot was sometimes hard put to hold T.R.'s enthusiasm in check. In January 1905, when 2,000 conservationists, foresters and lumbermen met in Washington to discuss their woodland interests, the President made a strong effort to draw the timbermen into partnership with the government in conservation work. Following the script he had prepared with Pinchot, T.R. disavowed political meddling in forest policy: "Henceforth the movement for the conservative use of the forest is to come mainly from within, and not from without; from the men who are actively interested in the use of the forest." But, having gotten off on the right foot, Roosevelt then proceeded to put his left foot squarely in his mouth. Departing from his text, he roundly lambasted lumbermen who "skin the country and go somewhere else . . . whose idea of developing the country is to cut every stick of timber off of

Titled *Views of the South Pacific Coast Railroad,* this 1890s lithograph shows the scenery Californians could enjoy along the line's

NATURAL BRIDGE SANTA CRUZ

GLIMPSE OF SANTA CRUZ FROM THE SAN LORENZO VALLEY

E AND THE SAN LORENZO

THE SAN LORENZO FROM THE R.R. CROSSING, BIG TREE STATION.

route between Santa Cruz and Oakland. During the week, the railroad primarily hauled redwood logs; weekends it catered to sightseers.

it, and leave a barren desert for the homemaker who comes in after him." T.R.'s tactlessness dealt a setback to government-industry cooperation.

A month later, Pinchot and Roosevelt finally won their battle to bring all government forest activities under Pinchot's control. By Act of Congress, the reserves were transferred from the Department of the Interior to the Department of Agriculture; Pinchot's division, renamed the United States Forest Service, was charged with protecting and maintaining them. From Secretary of Agriculture James Wilson, Pinchot received a sweeping franchise to act by the moderate principles he had always espoused: "All the resources of the Forest Reserves are for use and this must be brought about in a thoroughly prompt and businesslike manner, under such restrictions only as will insure the permanence of these resources. Where conflicting interests must be reconciled, the question will always be decided from the standpoint of the greatest good of the greatest number in the long run."

Pinchot's successful struggle to centralize government efforts in the forestry field was not his last battle —far from it. Every year he had to fight for the life of his service when appropriations bills came up in Congress. He still had to contend with intransigence from old-line loggers and millmen who saw his ideas as government intervention in their affairs. And they and their lobbyists and friends in Congress never missed an opportunity to attempt to throttle the Forest Service—to the point where President Roosevelt said of the service that "more time appeared to be spent upon it during the passage of appropriations bills than on all the other government bureaus put together."

Yet as time went on, more and more lumbermen came to realize that they could trust Pinchot and live with his pragmatic approach to conservation. It was by no means certain that lumbermen would adopt all the reforms that Pinchot advocated, but at least they could explore the possibilities without fear that the government was their enemy. Already Pinchot had seen encouraging shows of interest from the bellwether of the lumber industry—the Weyerhaeuser Timber Company.

The old baron's sons had become increasingly active in practical experiments. In 1903 they donated to the state university of Minnesota 2,640 acres of timberland as a test forest. The same year they invited Pin-

chot to send a forester to Minnesota to analyze the cost of raising a crop of trees on cut-over acreage. The forester, C. S. Chapman, conducted lengthy studies and concluded that after 50 years a new crop of white pine might pay for itself, but to achieve that, selective cutting would have to be done under trained foresters, and strict fire control would have to be enforced. The cost would be high, and the profit small at best.

Chapman's report confirmed the Weyerhaeusers' own observations—and their father's skepticism. They believed that lumbermen could afford to practice forest conservation only with government aid. Weyerhaeusers' West Coast manager, George Long, summed up their views when he told a reporter that if the forests were to be replaced, the state must either buy cut-over lands from the lumber companies and replant the forests, or remit the taxes on tracts replanted by the companies. Though such arrangements were not impossible, they would take years to develop.

Nevertheless, the Weyerhaeuser sons kept an open mind. At the forestry conference of 1905, Fred listened with apparent approval to the reading of a paper with a significant title: "The Changed Attitude of Lumbermen Toward Forestry." It was by J. E. Defebaugh, editor of *American Lumberman;* among other things, it expressed his pleasure that the conferees were a mixture of lumbermen, conservationists and government officials, all working for the common good. By way of proving the point, Fred Weyerhaeuser headed a lumbermen's committee that raised $100,000 to endow a new chair of forestry at Yale.

Enlightened self-interest prompted other loggers and millmen to follow the Weyerhaeusers' lead. Soon Pinchot-trained foresters were working in the woodlands on logging companies' payrolls, and young lumbermen of the big-business generation were discussing conservation measures at social functions Pinchot held at his home in Washington, D.C.

By 1910, conservation was more than a cause—it was a working reality. It had been made possible by American resources as abundant as the forests of the West: the tempered idealism of men like Gifford Pinchot and Theodore Roosevelt; the stamina and ingenuity of researchers in universities and foresters in the fields; a sense of public responsibility on the part of the press and public-minded lumbermen like the Wey-

Striking loggers in California's Converse Basin surround a sign proclaiming one of their grievances: a lack of sugar. A wage demand may have figured in the strike, but food was always a vital issue in the camps.

erhaeusers. It would take time to put into general practice the difficult techniques of selective cutting and sustained yield, but a start had been made.

As the basic work of fire prevention and forest management went forward, many conservationists learned to their surprise of a powerful ally: the incredible persistence of the forests themselves. In 1910, California millmen were producing lumber from second-growth redwood forests with trees 100 to 150 feet tall that had sprung up since their grandfathers clear-cut the area 60 years earlier. Those first loggers had no intention of sparing a tree, but they had left standing a few worthless redwoods—trees too old or flawed to yield salable lumber—and seeds from those lonely survivors had sired a new generation. It was enough to make any logger think wistfully of the vast forests that could be harvested if people actually helped them grow. And old John Muir, who had known for decades of the forests' powers of self-renewal, was banking on that factor when he predicted: "Our forests under rational management will yield a perennial supply of lumber for

every right use without further diminishing their area."

If anyone said the last word on the subject of conservation, it was Muir, who had devoted his entire life to a counterassault on man's destruction of the wilderness. He wanted to preserve the forests not merely as tree farms, which were nourished to supply man's physical needs, but as permanent wilderness that nourished man's spiritual needs. John Muir believed that everyone and everything was "hitched to everything else in the universe," and that one could not realize it fully until he stood in awe of a giant sequoia, walked in perfect solitude down a towering colonnade of Douglas firs, watched a tiny bird make its living on the tough little seeds of a western hemlock. As far back as 1869, when the naturalist first roamed through the evergreen uplands of the Sierra Nevada, he had written that trees were "the very gods of the plant kingdom, living their sublime century lives in sight of Heaven, watched and loved and admired from generation to generation! How rich our inheritance in these blessed mountains, the tree pastures into which our eyes are turned!"

225

The toughest railroading in the world

The Western lumber industry boomed from its very start, but it reached greatness with the development of the logging railroad. Once lumbermen moved into the mountains, away from rivers capable of floating their product to the mills, only a railroad could carry the quantities of huge logs they produced.

A logging railroad did not look very much like the Union Pacific. The tracks were light rails fastened to often flimsy ties on poorly ballasted roadbeds, and many of the locomotives —graced with such names as Coffee Grinder and Old Blue—resembled contraptions put together from junk-yard parts. But, operated by woodsmen-engineers, they climbed unbelievable grades, clattered across dizzyingly high log trestles and snaked loads of 80-foot logs around 30° curves in feats of railroading never seen before or since.

Having loaded a train of flatcars, loggers stand proudly by the donkey that hauled in the logs.

Moving through a lane of towering trees above Oregon's Santiam River valley, a trainload of logs heads for the Hammond Lumber Company mill at Mill City. Such single-track roads usually handled several trains at the same time, often without the benefit of signals or dispatchers.

A Shay locomotive, placed in the middle of the train to distribute the load evenly, pauses on a 100-foot-high wooden trestle across a Washington ravine. Flatcars were frequently dispensed with—as in three of the four sets of logs here; the wheel trucks were attached directly to the load.

Members of a woods crew of the Mendocino Lumber Company in California — some seated on logs left aboard for that purpose — crowd a flatcar used to transport them to and from camp. Some companies had more elaborate facilities: passenger cars bought secondhand from commercial railroads.

232

TEXT CREDITS

For full reference on specific page credits see bibliography.

Chapter 1: Particularly useful sources for information and quotes in this chapter were: Ralph Clement Bryant, *Logging,* John Wiley & Sons, 1913; Thomas R. Cox, *Mills and Markets,* University of Washington Press, 1974; William M. Harlow and Ellwood S. Harrar, *Textbook of Dendrology,* McGraw-Hill Book Co., 1969; Stewart Holbrook, *Holy Old Mackinaw,* The Macmillan Company, 1938; Edmond S. Meany, *Vancouver's Discovery of Puget Sound,* Binfords & Mort, Publishers, 1957; Donald Culross Peattie, *A Natural History of Western Trees,* Houghton Mifflin Co., 1953; 19 — Wilkeson quote, *Wilkeson's notes,* pp. 17-18. Chapter 2: Particularly useful sources for information and quotes. Edwin T Coman Jr. and Helen M. Gibbs, *Time, Tide and Timber: A Century of Pope & Talbot,* Greenwood Press, 1968; Thomas R. Cox, *Mills and Markets: A History of the Pacific Coast Lumber Industry to 1900,* University of Washington Press, 1974; John Robert Finger, *Henry L. Yesler's Seattle Years, 1852-1892,* University of Washington Ph.D. dissertation, 1968. Chapter 3: Particularly useful sources for information and quotes: Kramer Adams, *Logging Railroads of the West,* Bonanza Books, 1961; Ralph W. Andrews, *Redwood Classic,* Superior Publishing Co., 1958; Ralph Clement Bryant, *Logging,* John Wiley & Sons, 1913; Ruby El Hult, *Steamboats in the Timber,* Binfords & Mort, Publishers, 1952; Hank Johnston, *They Felled the Redwoods,* Trans-Anglo Books, 1966; H. S. Lyman, "Reminiscences of Clement Adams Bradbury, 1846," *Oregon Historical Quarterly,* Vol. 2, No. 3, September 1901; Walter F. McCulloch, *Woods Words: A Comprehensive Dictionary of Loggers Terms,* Oregon Historical Society and The Champoeg Press, 1958; Rev. Andrew Mason Prouty, *Logging with Steam in the Pacific Northwest, the Men, the Camps, and the Accidents 1885-1918,* Master's thesis, Department of History, University of Washington, 1973; Peter J. Rutledge and Richard H. Tooker, "Steam Power for Loggers: Two Views of the Dolbeer Donkey," *Forest History,* Vol. 14, No. 1, April 1970; 91 — Humboldt County logger quote: Andrews, *Redwood Classic,* p. 48; 96 — Engstrom quote: Engstrom, *Vanishing Logger,* pp. 7-9; 105 — Millard quotes: *Everybody's* magazine, Vol. 9, No. 1, July 1903; 116 — Paine quotes: *Outing Magazine,* September 1906. Chapter 4: Particularly useful sources for information and quotes: Ralph Hidy, Frank Ernest Hill and Allan Nevins, *Timber and Men: The Weyerhaeuser Story,* The Macmillan Company, 1963; Stewart Holbrook, *Green Commonwealth,* copyright Simpson Logging Company, printed by Frank McCaffrey; 138 — worker quote: *Lane County Historian,* Vol. 20, No. 1, Spring 1975; Oscar Page quote: written for Alfred D. Collier, Collier State Park Logging Museum; 146 — Florence Hills quote: Unpublished diary of 1905 of Florence Elizabeth Hills. Collection of her daughter, Hallie Huntington. Chapter 5: Particularly useful sources for information and quotes: Stewart H. Holbrook, *Far Corner: A Personal View of the Pacific Northwest,* The Macmillan Company, 1952; Stewart H. Holbrook, *Holy Old Mackinaw,* Ballantine Books, 1971; Ellis Lucia, *Head Rig: Story of the West Coast Lumber Industry,* Overland West Press, 1965; Murray Morgan, *The Last Wilderness,* Viking, 1955; Murray Morgan, *Skid Road,* Viking, 1951; 179 — Olsen quote on Erickson's: Andrews, *This Was Sawmilling,* p. 172; 180 — Humboldt Saloon quote: Andrews, *This Was Logging,* pp. 112 and 114. Chapter 6: Particularly useful sources for information and quotes: Laurence (Scoop) Beal, *The Carson Mansion,* Times Printing Co., 1973; Michael Frome, *Whose Woods These Are,* Doubleday, 1962; Ralph Hidy, Frank Ernest Hill and Allan Nevins, *Timber and Men: The Weyerhaeuser Story,* The Macmillan Company, 1963; Thomas Emerson Ripley, *Green Timber,* American West Publishing Company, 1968.

PICTURE CREDITS

The sources for the illustrations in this book are shown below. Credits from left to right are separated by semicolons, from top to bottom by dashes.

Cover — Courtesy The Bettmann Archive. 2 — George Weister, courtesy University of Oregon Library. 6,7 — Carleton E. Watkins, courtesy The Bancroft Library. 8,9 — Benjamin Gifford, courtesy University of Oregon Library. 10,11 — H. E. French, courtesy National Archives, #95-G-85562. 12,13 — A. W. Ericson, courtesy Humboldt State University Library. 14,15 — Courtesy Oregon Historical Society. 16,17 — C. L. Goddard, courtesy The Bancroft Library. 18 — A. W. Ericson, courtesy Humboldt State University Library. 20,21 — Map by Rafael D. Palacios. 23 — Courtesy General Research and Humanities Division, The New York Public Library, Astor, Lenox and Tilden Foundations; Derek Bayes, courtesy Linnean Society of London. 24,25 — Drawings by Don Bolognese — Frank Lerner, courtesy Library of the New York Botanical Garden. 26,27 — Drawings by Don Bolognese — Frank Lerner, courtesy Library of the New York Botanical Garden. 28,29 — Derek Bayes, courtesy The National Maritime Museum, London, on loan from Ministry of Defense (Navy). 32 — Bertram Buxton, courtesy of the British Columbia Provincial Museum, Victoria, British Columbia. 34,35 — Constance Gordon-Cumming, *The Father of the Forest (The Calaveras Grove)* California, 1878. The Kahn Collection, The Oakland Museum, Oakland, California. Copied by Dean Austin. 37 — Courtesy Library of Congress. 40,41 — H. C. Tibbits, courtesy The Bancroft Library. 42,43 — Courtesy The Bancroft Library. 44,45 — A. R. Moore, courtesy Harold G. Schutt. 46,47 — A. W. Ericson, courtesy Peter Palmquist Collection. 48,49 — Jay Golden, courtesy Mary Roberts Horton. 50 — Courtesy Union Lumber Co. Collection, San Francisco Maritime Museum. 52 — George H. Knight, courtesy Society of California Pioneers, San Francisco, Calif. 53 — Courtesy The Bancroft Library. 56 — Courtesy Pope & Talbot, Inc. 59 — Ellis Herwig, courtesy East Machias Public Library. 60,61 — Floyd Lee, courtesy Seattle Historical Society. 62 — Courtesy Pope & Talbot, Inc. 65 — Floyd Lee, courtesy Seattle Historical Society. 66 — Courtesy Del Norte County Historical Society, Crescent City, Calif. 68,69 — Floyd Lee, courtesy Henry Leiter Yesler Papers, University of Washington Library — Courtesy Seattle Historical Society; inset top right, Asahel Curtis, courtesy Seattle Public Library. 70,71 — Courtesy of The California Historical Society, San Francisco/San Marino. 72,73 — Courtesy San Francisco Maritime Museum. 74,75 — Courtesy Collection of Robert J. Lee. 76,77 — Courtesy Jack Lowe Collection, San Francisco Maritime Museum. 78,79 — Hazeltine, courtesy The Bancroft Library. 80,81 — Darius Kinsey, courtesy The Darius Kinsey Collection. 82,83 — Courtesy National Archives, #95-G-32573. 84,85 — Courtesy Oregon Historical Society. 86,87 — Courtesy Harold G. Schutt. 88 — Courtesy Collection of Robert J. Lee. 91 — Courtesy Hank Johnston Collection. 92,93 — John Zimmerman, courtesy Fort Humboldt State Historic Park, California De-

ACKNOWLEDGMENTS

The editors wish to give special thanks to Ronald Fahl and Dr. Harold Steen, Forest History Society, Santa Cruz, Calif., and Professor Thomas R. Cox, Department of History, San Diego State University, who read and commented on major portions of the book.

The editors also wish to thank: George Abdill, Douglas County Museum; Pamela Allsebrook, California Redwood Association; Rangers Jim Anderson and John Knot, Fort Humboldt State Historic Park, California Dept. of Parks and Recreation; Marjorie Arkelian, The Oakland Museum; Curt Beckham, Myrtle Point, Ore.; Stephen Dow Beckham, Linfield College; Dave Bohn, The Darius Kinsey Collection, Berkeley, Calif.; Lee Burtis, Catherine Hoover, Gary Kurutz, Maude K. Swingle, California Historical Society; Lynwood Carranco, Arcata, Calif.; Frank Colburn, Georgia Pacific Museum; Alfred D. Collier, Collier State Park Logging Museum; Gary Cook, Pacific Lumber Company; Chris Corbin, The Ingomar Club, Eureka; James H. Davis, Idaho State Historical Society; J. Robert Davison, Provincial Archives of British Columbia; Dr. Carl S. Dentzel, Southwest Museum; Matilda Dring, San Francisco Maritime Museum; B. Shirley Edwards, Pauline Lehman, Del Norte County Historical Society; Richard H. Engeman, Janice Worden, Dan Nielsen, Oregon Historical Society; Emory Escola, Little River, Calif.; Nannie M. Escola, Mendocino, Calif.; David Featherstone, Martin Schmitt, Carrie Singleton, Univ. of Oregon Library; Caroline Gallacci, Frank Green, Washington State Historical Society; Suzanne H. Gallup, William Roberts, The Bancroft Library; Andrew Genzoli, Ferndale, Calif.; Jay Golden, The Photo Lab Inc., Roseburg; Kathleen Grasing, Oregon State Library; Barbara Gordon, Forestry Library, Univ. of Washington; Ronald Grefe, Patent Reproduction Company, Washington, D.C.; Jack Gyer, National Park Service, Yosemite National Park, Calif.; Villette Harris, Washington, D.C.; W. C. Haskins, Mel Von Busch, Simonds Cutting Tools, Wallace Murray Corp.; Finley Hayes, *Loggers' World;* Sally Heick, Patricia Hutcherson, Pope and Talbot, Seattle; Hallie Huntington, Eugene, Ore.; H. R. Hutchins, Pope and Talbot, Portland; L. James Higgins Jr., Nevada State Historical Society; Mary Roberts Horton, Roseburg; David James, Simpson Timber Co.; Forrest and Merton Johnson, Ellensburg, Wash.; Alan Jutzi, Henry E. Huntington Library; John T. Labbe, Beaverton, Ore.; William Leary, Douglas Thurman, National Archives, Washington, D.C.; Robert J. Lee, Ukiah, Calif.; Robert F. Lussier, E. W. Nolan, Bertha Stratford, Seattle Historical Society; Paula Maker, East Machias Public Library; Glenn Mason, Lane County Pioneer Museum; Arthur J. McCourt, Frances Connors, Historical Archives, Weyerhaeuser Company, James H. Mitchell, Georgia Pacific Corp., Fort Bragg; Elwood Mounder, Forest History Society; Guy Page, Waldport, Ore.; Peter E. Palmquist, Arcata; Charles Peck, Pope and Talbot, Port Gamble, Wash.; Mrs. Guy B. Pope, Portland; Reverend Andrew Mason Prouty, Auburn, Wash.; Virginia Reich, Seattle Public Library; J. Roger-Jobson, Mrs. Irene Lichens, The Society of California Pioneers; Erich Schimps, Humboldt State Univ. Library; Harold G. Schutt, Lindsay, Calif.; Diane Schwartz, Library of the New York Botanical Garden; Pat Stafford, Burlington Northern; Kay Terry, Carol Ventgen, City of Coos Bay Public Library; Henry Wangeman, Ellensburg, Wash.; William Wilson, Coeur d'Alene Public Library; Karyl Winn, Carol Zabilski, Suzzallo Library, Univ. of Washington.

BIBLIOGRAPHY

Abdill, George B., *Pacific Slope Railroads.* Superior, 1959.

Adams, Kramer, *Logging Railroads of the West.* Superior, 1961.

Allen, Alice Benson, *Simon Benson, Northwest Lumber King.* Binfords & Mort, 1971.

Allen, James B., *The Company Town in the American West.* University of Oklahoma Press, 1967.

Andrews, Ralph W.:
Glory Days of Logging. Superior, 1961.
Heroes of the Western Woods. E. P. Dutton, 1960.
Redwood Classic. Superior, 1968.
This Was Logging! Superior, 1954.
This Was Sawmilling. Bonanza Books, 1957.
Timber. Superior, 1968.

Bagley, Clarence Booth:
The History of Seattle, Vol. 1, 1916.
" 'The Mercer Immigration': Two Cargoes of Maidens for the Sound Country." *Oregon Historical Quarterly,* Vol. 5, No. 1, March 1904.

Beale, Laurence (Scoop), *The Carson Mansion.* Times Printing Company, 1973.

Beckham, Curt, "The Disaster at Trestle Creek." *The Westerner,* January-February 1974.

Beckham, Stephen Dow, *Coos Bay: The Pioneer Period, 1851-1890.* Arago Books, 1973.

Binns, Archie:
Northwest Gateway. Doubleday, Doran & Company, Inc., 1941.
The Roaring Land. Robert M. McBride & Company, 1942.

Bohn, Dave, and Rodolfo Petschek, *Kinsey, Photographer,* Vol. I.

Scrimshaw Press, 1975.

Bryant, Ralph Clement, *Logging.* John Wiley & Sons, 1913.

Carranco, Lynwood, and John T. Labbe, *Logging in the Redwoods.* The Caxton Printers, Ltd., 1975.

Clar, C. Raymond:
"John Sutter, Lumberman." *Journal of Forestry,* Vol. 56, No. 4, April 1958.
"The Spurious Collateral of Honest Harry Meiggs." Written for *California Historical Society Quarterly* but apparently unpublished.

Clark, Donald H., "Sawmill on the Columbia." *The Beaver,* June 1950.

Coman, Edwin T., Jr., and Helen M. Gibbs, *Time, Tide and Timber: A Century of Pope & Talbot.* Greenwood Press, 1968.

Cox, Thomas R.:
Mills and Markets. University of Washington Press, 1974.
"Lumber & Ships: The Business Empire of Asa Mead Simpson." *Forest History,* Vol. 14, No. 2, July 1970.

Dana, Julian, *Sutter of California.* Press of the Pioneers, 1934.

Dawdy, Doris O., *Artists of the American West.* Swallow, 1974.

Denny, Emily Inez, *Blazing the Way.* Seattle, 1909.

Eastwood, Alice, "Early Botanical Explorers on the Pacific Coast and the Trees They Found There." *California Historical Society Quarterly,* Vol. 18, No. 4, December 1939.

Elliott, Eugene Clinton, *A History of Variety-Vaudeville in Seattle from the Beginning to 1914.* University of Washington Press, 1944.

Engstrom, Emil, *The Vanishing Logger.* Vantage Press, 1956.

Finger, John R.:
Henry L. Yesler's Seattle Years, 1852-1892. University of Wash-

ington Ph.D. thesis, 1968.

"Henry Yesler's 'Grand Lottery of Washington Territory.'" *Pacific Northwest Quarterly,* July 1969.

"Seattle's First Sawmill 1853-1869." *Forest History,* January 1972.

Frome, Michael, *Whose Woods These Are, the Story of the National Forests.* Doubleday & Company, 1962.

Gay, Theressa, *James W. Marshall: The Discoverer of California Gold.* The Talisman Press, 1967.

Genzoli, Andrew, *Samoa Cookhouse Memories.*

Harker, Mary Margaret, "'Honest Harry' Meiggs." *California Historical Society Quarterly,* Vol. 17, No. 3, September 1938.

Harlow, William M., and Ellwood S. Harrar, *Textbook of Dendrology.* McGraw-Hill Book Company, 1969.

Hidy, Ralph W., Frank Ernest Hill and Allan Nevins:
Timber and Men, The Weyerhaeuser Story. The Macmillan Company, 1963.
History of Mendocino County. Alley, Bowen Company, 1880.

Hittel, John S., *A History of the City of San Francisco.* A. L. Bancroft & Company, San Francisco, 1878.

Hoffman, Daniel G., *Paul Bunyan.* University of Pennsylvania Press for Temple University Publications, 1952.

Holbrook, Stewart H.:
Far Corner. The Macmillan Company, 1952.
Half Century in the Timber. Frank McCaffrey, 1945.
Holy Old Mackinaw. The Macmillan Company, 1938.
The Yankee Exodus. The Macmillan Company, 1950.
"Daylight in the Swamp." *American Heritage,* October 1958.

The Home of the Redwood. The Redwood Mfg. Association Pacific Coast Wood and Iron, 1897.

Hult, Ruby El, *Steamboats in the Timber.* Binfords & Mort, 1952.

Hutchinson, W. H., *California Heritage: A History of Northern California Lumbering.* Diamond Gardner Corp., 1958.

Ingersoll, Ernest, "In a Redwood Logging Camp." *Harper's New Monthly Magazine,* January 1883.

Jackson, W. A., *The Doghole Schooners.* California Traveler, 1969.

James, Dave, *Big Skookum.* Shelton, Washington, 1953.

Johnson, Paul C., *Sierra Album.* Doubleday & Company, Inc., 1971.

Johnston, Hank, *They Felled the Redwoods.* Trans-Anglo, 1966.

Keithahn, Edward L., *Monuments in Cedar.* Bonanza Books, 1963.

Kortum, Karl, and Roger Olmsted, *Sailing Days on the Redwood Coast.* California Historical Society, 1971.

Kracht, Shannon, "Wendling, a Company Town." *Lane County Historian,* Vol. 20, No. 1, Spring 1975. Lane County Historical Soc.

Labbe, John T., and Vernon Goe, *Railroads in the Woods.* Howell-North, 1961.

Lockley, Fred, "Grays Harbor, The Largest Lumber-Shipping Port in the World." *The Pacific Monthly,* June 1907.

Lucia, Ellis:
The Big Woods. Doubleday & Company, Inc., 1975.
Head Rig: Story of the West Coast Lumber Industry. Overland West Press, 1965.

"Lumbering in Washington Territory." *The Overland Monthly,* Vol. 5, July 1870. John H. Carmany and Company.

Lyman, H.S., "Reminiscences of Clement Adams Bradbury, 1846." *Oregon Historical Quarterly,* Vol. 2, No. 3, September 1901.

McCulloch, Walter F., *Woods Words: A Comprehensive Dictionary of Loggers Terms.* The Oregon Historical Society and the Champoeg Press, 1958.

McNairn, Jack, and Jerry MacMullen, *Ships of the Redwood Coast.* Stanford University Press, 1945.

Mason, Glenn, "River Driving in Lane County." *Lane County Historian,* Vol. 18, No. 2, Summer 1973.

Meany, Edmond S., *Vancouver's Discovery of Puget Sound.* Binfords & Mort, 1957.

Melendy, H. Brett, "Two Men and a Mill." *California Historical Society Quarterly,* Vol. 38, March 1959.

Millard, Bailey, "Flying Down a Fifty-Mile Flume." *Everybody's Magazine,* Vol. 9, No. 1, July 1903.

Morgan, Murray:
The Last Wilderness. The Viking Press, 1955.
The Northwest Corner. The Viking Press, 1962.
Skid Road: An Informal Portrait of Seattle. The Viking Press, 1951.

Moungovan, Thomas O., Julia L. Moungovan and Nannie Escola, *Logging with Ox Teams: An Epoch in Ingenuity.* Mendocino County Historical Society, 1968.

Paine, Ralph D., "The Builders: The Heart of the Big Timber Country." *Outing Magazine,* September 1906.

Palmquist, Peter E., *Fine California Views: The Photographs of A. W. Ericson.* Interface California Corp., 1975.

Peattie, Donald Culross, *A Natural History of Western Trees.* Houghton Mifflin Co., 1953.

Phelps, Thomas Stowell, Commodore U.S.N., *Reminiscences of Seattle, Washington Territory and the U.S. Sloop-of-War Decatur during the Indian War of 1855-56.* L. R. Hamersly and Co., 1881.

Prouty, Father Andrew Mason, *Logging with Steam in the Pacific Northwest: The Men, The Camps, and The Accidents 1885-1918.* M.A. thesis. Dept. of History, Univ. of Washington, 1973.

Ripley, Thomas Emerson, *Green Timber.* American West, 1968.

Rutledge, Peter J., and Richard H. Tooker, "Steam Power for Loggers: Two Views of the Dolbeer Donkey." *Forest History,* Vol. 14, No. 1, April 1970.

Ryder, David Warren, *Memories of the Mendocino Coast.* Privately printed by Taylor & Taylor, 1948.

Sargent, Charles Sprague, *Silva of North America,* Vols. 10, 11, 12. Houghton, Mifflin Co., 1898.

Smith, David C., "The Logging Frontier." *Forest History,* Vol. 18, No. 4, October 1974.

Speidel, William C., *Sons of the Profits.* Nettle Creek, 1967.

Stevens, James:
Paul Bunyan. Garden City Publishing Co., Inc., 1925.
Timber! The Way of Life in the Lumber Camps. Row, Peterson, 1942.

Stewart, Watt, *Henry Meiggs: Yankee Pizarro.* AMS Press, 1968.

Tooker, R. H., "Loading by High-Line on the California Coast." *Sea Letter,* Vol. 6, No. 1, May 1968, San Francisco Maritime Museum.

Untermeyer, Louis, *The Wonderful Adventures of Paul Bunyan.* The Heritage Press, 1945.

Wadsworth, Wallace, *Paul Bunyan 'and His Great Blue Ox.'* Doubleday & Company, Inc., 1926.

Warren, Jim, *Pictorial History of Seattle,* 1964.

Weinstein, Robert A., "Lumber Ships at Puget Sound, a Photographic Record by Wilhelm Hester." *American West Magazine,* Vol. 2, No. 2, Fall 1965.

Wilkeson's Notes on Puget Sound. Being Extracts from Notes by Samuel Wilkeson of a Reconnaissance of the Proposed Route of the Northern Pacific Railroad made in the summer of 1869.

INDEX *Numerals in italics indicate an illustration of the subject mentioned.*